Information Retrieval and Documentation in Chemistry

CONTRIBUTIONS IN LIBRARIANSHIP AND INFORMATION SCIENCE

Series Editor: *Paul Wasserman*

1. Urban Analysis for Branch Library System Planning
   Robert E. Coughlin, Françoise Taieb, and Benjamin H. Stevens
2. Frontiers in Librarianship: Proceedings of the Change Institute, 1969
   School of Library and Information Services, University of Maryland
3. Subject Retrieval in the Seventies: New Directions. An International Symposium
   Hans (Hanan) Wellisch and Thomas D. Wilson, Editors
4. Quantitative Methods in Librarianship: Standards, Research, Management
   Irene Braden Hoadley and Alice S. Clark, Editors
5. Public Relations for Libraries: Essays in Communications Techniques
   Allan Angoff, Editor
6. Human Memory and Knowledge: A Systems Approach
   Glynn Harmon
7. Libraries in the Political Scene: Georg Leyh and German Librarianship, 1933-1953
   Marta L. Dosa

Information Retrieval and Documentation in Chemistry

CHARLES H. DAVIS and JAMES E. RUSH

Contributions in Librarianship and Information Science, Number 8

GREENWOOD PRESS

Westport, Connecticut • London, England

Library of Congress Cataloging in Publication Data

Davis, Charles Hargis, 1938-
Information retrieval and documentation in chemistry.

(Contributions in librarianship and information science, no. 8)
Includes bibliographical references.
1. Information storage and retrieval systems—Chemistry. I. Rush, James E., joint author. II. Title. III. Series [DNLM: 1. Chemistry. 2. Information retrieval systems. Z699.5C5 D26li 1974]
Z699.5.C5D38 029'.9'54 72-791
ISBN 0-8371-6364-1

Library of Congress Catalog Card Number: 72-791

ISBN: 0-8371-6364-1

First published in 1974

Greenwood Press, a division of Williamhouse-Regency Inc.
51 Riverside Avenue, Westport, Connecticut 06880

Manufactured in the United States of America

To

Bonnie and Dee

Contents

Figures

Tables

Preface

For several years we have thought that greater effort should be expended to create more understanding of the processes involved in information storage, retrieval, and dissemination. Too often, concepts deriving from science and technology are made to seem unnecessarily abstruse, either inadvertently or deliberately to preserve some sort of mystique. This book is intended to explicate, if not popularize, major aspects of I S & R processes as they are exemplified by the field of chemical documentation. To the extent that this work is found useful by chemists, information specialists, and all individuals interested in scientific documentation, we shall be gratified.

The material is organized into nine chapters. The first six may be read and understood without any special knowledge of chemistry. Following a brief introduction in Chapter 1, Chapter 2 examines the important but often neglected human factor. Since I S & R systems are designed by humans for use by humans, this factor surely deserves first consideration.

Chapter 3 treats the subject of indexing in a very broad sense, and covers subject and keyword indexing, classification, and so on. A number of specific indexes are discussed and illustrated.

Chapter 4 discusses abstracting, with emphasis on automatic abstracting.

Chapter 5 deals with retrieval techniques and systems. The basics of Boolean and weighted-term searches are provided, and factors such as truncation are discussed.

Chapter 6 focuses on the difficult subject of evaluation. The subject is difficult because everyone has an opinion as to the worth of something, thereby making objectivity an elusive commodity. For this reason the factors involved in evaluation are delineated as clearly as possible in this chapter.

The remaining three chapters center on topics that, on the surface, seem specific to chemistry. For instance, Chapter 7 treats chemical nomenclature. But it will be seen that chemical names are just linear representations of nonlinear things and bear close similarities to the nomenclatures of other fields. For the nonchemist, perhaps the lesson to be learned from this chapter is that the naming process is important, dependent upon the application of a name, and closely resembles both classification schemes and languages.

Chapter 8 examines the forms of nomenclature that are not dealt with in Chapter 7. The development and application of these nomenclatures (notations) have received a great deal of attention from many people; accordingly, the scope of the topic is scarcely even hinted at in this chapter. Nevertheless, notation development has many parallels in other disciplines, such as music, librarianship, and mathematics.

Finally, Chapter 9 discusses a topic which in many respects belongs to the field of graph theory as much as to chemistry. The material in this chapter is of historical, theoretical, and practical importance in chemistry. Some of the concepts are also of considerable mathematical consequence. An additional purpose of this chapter is to introduce the reader (by example) to the notion of algorithms and to some of the rudiments of file design. Many references are included to lead the reader to more detailed works on the material discussed.

The reader is encouraged in reading this work to try to avoid the narrowness of mind which discipline-oriented works often foster. We have intended this book to appeal to a broader audience than that interested solely in chemical information. It remains to be seen how well we have succeeded.

Charles H. Davis and James E. Rush
Drummond Island, Michigan
May 1973

Acknowledgments

All artwork has been prepared by the authors. The chemical structures were done on a DURA MACH-10 typewriter, using the typing element developed at Smith, Kline & French Laboratories, Philadelphia, Pennsylvania. Photoprints of all art were prepared by Mr. Gene Hayner and his staff in Engineering Publications, Engineering Experiment Station, The Ohio State University. The final manuscript was typed by Mrs. Mary Kimball. Parts of the manuscript were read by a number of people, including Dr. Robert Fugmann, Dr. Carlos Bowman, Dr. Casimir Gladys, Mrs. Patricia Elliott, and Dr. Bertrand Landry. To all those who in any way contributed to the development of this book we say a heartfelt thanks.

All copyrighted material in this book is used with permission. We thank the publishers for allowing us to reprint the following artwork: Figure 6, from Frank Harary, *Proof Techniques in Graph Theory,* 1969, p. 263 (by permission of Academic Press, Inc., and the author); Figure 7, from *Dissertation Abstracts Retrospective Index,* Vol. II, *Chemistry,* 1970, p. 630 (by permission of University Microfilms); Figure 8, from Science Citation Index, 1972 (by permission of the Institute for Scientific Information); Figure 9(a), from Formula Index to Chemical Abstracts, Vol. 74, 1971, p. 202F (by permission of the

Chemical Abstracts Service); Figure 9(b), from *Beilsteins Handbuch der Organischen Chemie,* vierte auflage, band 29, zweiter teil, 1940, p. 1246 (by permission of Springer-Verlag); Figure 10(a), from the *HAIC Index to Chemical Abstracts,* Vol. 74, 1971, p. 273H (by permission of the Chemical Abstracts Service); Figure 10(b), from the *Rotaform Index,* 1972 annual (by permission of the Institute for Scientific Information); Figure 11, from the *Index of Ring Systems in Chemical Abstracts,* Vol. 73, 1970, p. 28R (by permission of the Chemical Abstracts Service); Figure 14, from *American Documentation,* Vol. 13, No. 4, 1962, pp. 378-390 (by permission of the American Society for Information Science); Figures 15 and 16, from H. P. Edmundson, "New Methods in Automatic Extracting," *Journal of the Association for Computing Machinery,* Vol. 16, No. 2, April 1969, pp. 264-285, Copyright 1969, Association for Computing Machinery, Inc. (by permission of the Association); Figure 17, from the *Journal of the American Society for Information Science,* Vol. 22, No. 4, July-August 1971, p. 260, Copyright 1971 by the American Society for Information Science, 1140 Connecticut Avenue, N.W., Washington, D.C. 20036 (by permission of ASIS); Figure 18, from *IFIP Congress* 71, Booklet TA-6, 1971, pp. 133-137 (by permission of North Holland Scientific Publishers); and Figure 25, from Charles P. Bourne, *Modern Methods of Information Handling,* 1963, p. 90 (by permission of John Wiley and Sons, Inc.).

Information Retrieval and Documentation in Chemistry

1

Introduction

In an attempt to emphasize both the general and the specific aspects of information retrieval and documentation in chemistry, this book has been organized in a modular fashion. Although they have chemical overtones, Chapters 2 through 6 represent a primer or introduction to the field of documentation in that they cover the human elements—indexing and classification, abstracting, principles of information retrieval, and the evaluation of information systems and services. These chapters may be profitably studied without knowledge of chemistry.

Chapters 7 through 9 treat nomenclature, linear notations, and topological systems for representing chemical structures. These are specific areas which will be of chief concern only to those with some background in chemistry. These chapters may be studied in isolation if the reader has adequate background in the other areas treated in this book.

FOR THE READER

Since information retrieval and documentation are part of the communication process, the reader may find it helpful in keeping his perspective if he has a general view of the process before pursuing the remainder of the book. A generalized block-diagram of the communication process is given in Figure 1. The reader can use the diagram as a "road-map" whenever he needs to know where he is with respect to the field as a whole.

SCOPE

It is sometimes useful to point out what a book is not, or at least to indicate those things which it does not attempt to cover. This book is *not* intended as a guide to the literature of chemistry. Adequate guides already exist, several of which are cited in Chapter 2. Moreover, the authors recognize that each of the chapters could be expanded to fill books in their own right; therefore, omissions have been inevitable. We have attempted to produce a work at the introductory to intermediate level which pulls together the major aspects of chemical documentation. More advanced or detailed works, selected on the basis of their pertinence and timeliness, are cited at the ends of the various chapters. Books dealing with science and technology are notorious for the speed with which they become obsolescent if not obsolete—the typical period of usefulness being from 5 to 10 years. Accordingly, recent publications may occasionally be cited in preference to more comprehensive but badly dated "classics."

OTHER PUBLICATIONS

Other works dealing with the topic of this book exist but are often out of date or consist merely of chapter-length reviews. Probably the best general review of work prior to 1967 is the chapter written by Tate for the second volume of the *Annual Review of Information Science and Technology.*[1] This contribution is distinguished by its excellent bibliography which contains 179 references broken down into six categories.

Another useful publication is an article by Garfield, et al., in the *Encyclopedia of Chemical Technology.*[2] It is less comprehensive than Tate's work but is more up to date. In addition, the National Academy of Sciences has published a review of the literature dealing with the

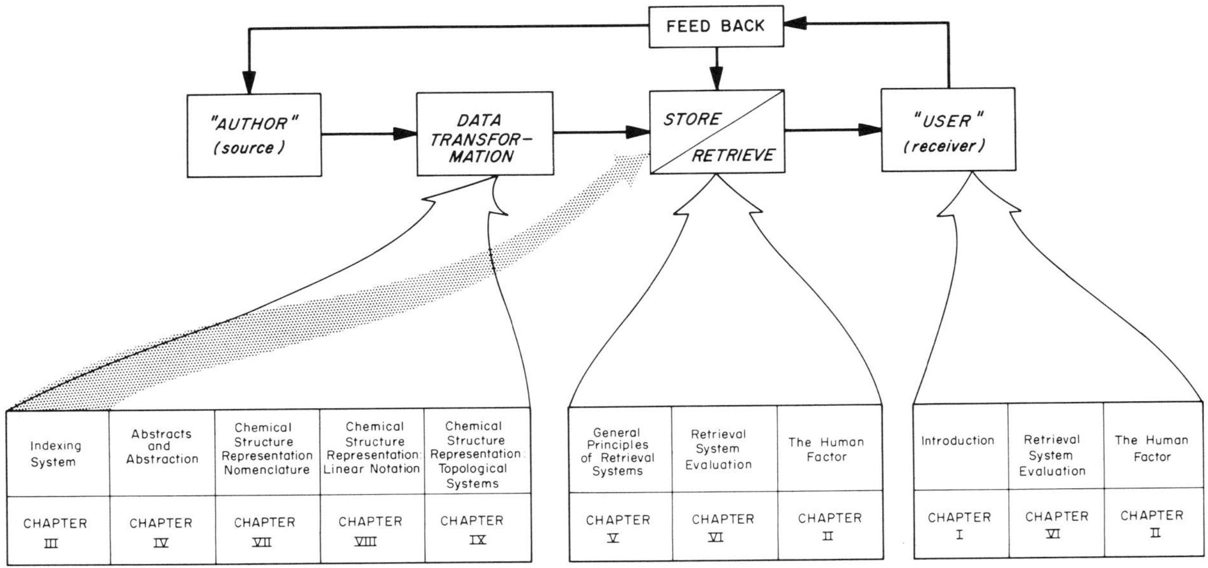

Figure 1—Relation Between the Parts of This Book and the General Communication Process

handling of chemical structure information, but this book is restricted to the period 1962 to 1968 and deliberately omits such major areas as chemical nomenclature.[3]

Books that are more than ten years old are probably of historical interest only. Two which fall into this category but still deserve mention because of their comprehensiveness are *Searching the Chemical Literature,*[4] published by the American Chemical Society in 1961, and the *Progress Report in Chemical Literature Retrieval,*[5] by Peakes, Kent, and Perry, published in 1957. Of course, some material, such as that by Vickery on classification and indexing in science, does not go out of date so quickly.[6]

A concise and useful book by Lynch, et al., appeared in 1971. This work treats computer handling of chemical structure information, but again traditional nomenclature is omitted.[7]

NOTES

1. Fred A. Tate, "Handling of Chemical Compounds in Information Systems," *Annual Review of Information Science and Technology,* Vol. 2 (New York: Wiley, 1967), pp. 285-309.

2. Eugene Garfield, Charles E. Granito, and Anthony E. Petrarca, "Information Retrieval Services and Methods," *Encyclopedia of Chemical Technology,* Supplement Volume, 2d ed. (New York: Wiley, 1971), pp. 510-535.

3. National Academy of Sciences, *Chemical Structure Information Handling: A Review of the Literature 1962-1968* (Washington, D.C.: 1969).

4. American Chemical Society, *Searching the Chemical Literature,* Advances in Chemistry Series Number 30 (Washington, D.C.: 1961).

5. Gilbert L. Peakes, Allen Kent, and James W. Perry, *Progress Report in Chemical Literature Retrieval* (New York: Interscience, 1957).

6. Brian C. Vickery, *Classification and Indexing in Science,* 2d ed. (New York: Academic Press, 1959).

7. Michael F. Lynch, Judith M. Harrison, and William G. Town, *Computer Handling of Chemical Structure Information* (New York: American Elsevier, 1971).

2

The Human Factor

An age of ever-increasing complexity requires increasingly complex information systems to facilitate communication. There are sometimes complaints, however, that too much emphasis is placed on technology at the expense of better service to the user. The computer, in particular, is often accused of having a "dehumanizing" effect on society.

It is undeniably true that information specialists may occasionally seem preoccupied with the mechanics of making a system more efficient or, for that matter, with making it work in the first place. However, it should also be obvious that the provision of better service to the user and the intelligent application of modern technology are not mutually exclusive enterprises, and that the computer, as yet unable to reply to its critics unassisted, is usually made the scapegoat when something goes wrong. In fact, if it were not for the computer, we would be living in a Kafkaesque nightmare, replete with error-prone and degrading tasks which humans would be forced to do. There is also the very real possibility that if we were still using strictly manual procedures we could no longer manage.

The point is, that no matter what devices are employed or how complicated an information system is, its basic purpose remains the same—to help people.

THE USER

Specialists in information science are quite concerned with investigating the needs and desires of users and potential users of information systems because these are of primary importance in systems analysis and design. It is not enough to make assumptions about the user or to do what seems intuitively correct. Ideally, information science studies will employ all the relevant techniques developed in the behavioral and social sciences.

Even the development of the realization that an information need exists is more subtle than one might think. For example, there may be as many as four different levels of question formation prior to the use of any information system:[1]

(1) The actual but unexpressed need for information (a visceral or subconscious need).

(2) Conceptualization of the need within the mind (a conscious need).

(3) A formal statement of the need in natural language (the formalized need).

(4) A statement of the need in the interface language, the language of the information system (the "compromised" need).

Once the need for information has reached the conscious level, an individual seeking to satisfy this need has several alternatives available. These options can be illustrated with a flow diagram,[2] or even by a simple hierarchy as shown in the following listing.

INFORMATION NEED
- I. Experiment
- II. Ask Colleague
- III. Search Literature
    - A. Personal Files
    - B. Library/Information Center
        - 1. Self Help
            - a. Search Strategy

2. Librarian/Information Specialist
    a. Question Negotiation
    b. Search Strategy

Initially, the individual has three options: (1) he may elect to satisfy his information need empirically by observing nature (for example, by determining a melting point in the laboratory); (2) he may simply ask somebody (usually a colleague) who is likely to know the answer to his question; or (3) he may choose to look the answer up. If a search of the literature is indicated, there are basically two possibilities: the person may look through the data that he already has at hand, or he may go to a library or information center. Depending on the complexity of the problem and the individual's experience with interrogating an information system, the person may then elect to do the search himself, or he may consult a reference librarian or information specialist. Should the person choose to do the work himself, then he must formulate his own strategy for interrogating the system. Otherwise, there will be an intermediate stage involving question negotiation between the user and the information specialist to determine precisely what the information need is. Following question negotiation, the information specialist usually formulates the search strategy, because of his greater familiarity with the system being used.

Naturally, if the query is of the type "What is the melting point of compound X?", there will not be much room for question negotiation. Information scientists therefore refer to this type of query as a "command" rather than a question.

The second principal means of satisfying an information need, the option of consulting a colleague, has proven to be of special importance. It seems that scientists who are on the frontiers of research, regardless of the field, tend to communicate directly with one another about their work. These researchers establish what have been dubbed "invisible colleges," and set up communication links that are independent of the literature of the fields in which they work. This phenomenon has been well described by Price:

> . . . one of the great consequences of the transition from Little Science to Big Science has been that after three centuries the role

> of the scientific paper has drastically changed. In many ways the modern ease of transportation and the affluence of the elite scientist have replaced what used to be effected by the publication of papers. We tend now to communicate person to person instead of paper to paper. In the most active areas we diffuse knowledge through collaboration. Through select groups we seek prestige and the recognition of ourselves by our peers as approved and worthy collaborating colleagues. We publish for the small group, forcing the pace as fast as it will go in a process that will force it harder yet. Only secondarily, with the inertia born of tradition, do we publish for the world at large.[3]

Of what use *is* the traditional literature, then, if it is dismissed in such a cavalier fashion by the members of the invisible colleges? There are several answers to this question. First, the members of invisible colleges constitute an elite within their own fields but not necessarily within other fields which may be of some importance. An expert in one field is often no better off than a layman in another, and his dependence on the literature becomes proportionately higher the farther he strays from his own area of expertise. This point has become more important in the latter part of the twentieth century, when the traditional distinctions between fields have become less clear and more and more interdisciplinary programs have been started. Second, the majority of the people working in a given field are not operating on the cutting edge of research. There are teachers, students, and practitioners who have perfectly legitimate informational needs which can be satisfied by sources that are slightly dated. Third, there is the question of accuracy. People engaged in research are constantly formulating and testing new hypotheses, and they are always looking for fresh ways of thinking about a subject. Consequently, their ideas may be stimulating but incorrect. The information exchanged by these people is different from that which most of the workers in the field need. The publication of articles in reputable journals is not just a technique for the dissemination of information; it is also a guarantee that the information contained in those articles is reasonably accurate.

In fact, the entire process of publication is more complex than many people realize. After a researcher initiates his work, he may make either verbal or written reports of his preliminary findings. Upon completion of the work, he may again make reports to informal audiences prior to submitting a manuscript to a journal or contributing a paper to convention proceedings. If the manuscript is rejected at this stage, it may be salvaged and published in a different journal at a later time. Next, the work will be picked up by an appropriate abstracting and indexing service and will be announced there. Further announcement will be made in annual reviews, the article will be cited in other journal articles, and eventually, usually after several years have elapsed, reference to the work will appear in various texts and treatises on the subject.

Some scientists consider this quality control feature of formal publication to be of preeminent importance. Garvey and Griffith are especially outspoken on this topic:

> . . . accelerating the flow of scientific information in the informal domain and expanding its dissemination is a problem precisely because it occurs in systems that obscure the boundary between the informal and formal domains. This boundary is one that science has deliberately erected to curtail, temporarily, the flow of information until the information has been examined against the current state of knowledge in a discipline. Non-scientists view procedure of curtailment as ultraconservative; experienced, practicing scientists perceive it as the essential feature of science. The scientist knows that at some point in this process, a research result, if it is ever to have relevance to science, must become eligible for status as a scientific finding, and a scientific finding, in turn, becomes genuine only if it can contribute to the understanding of a previously unexplained phenomenon. The long judicious procedure by which this conversion is made is unique to science. To reorganize it for the sake of speed, or for open communication with other spheres of intellectual endeavor, would almost certainly dismantle the institution of science as we know it today.[4]

There is definitely some merit in this position, although it is probably a bit overstated. Information systems should not be thought of merely as mechanisms for disseminating raw and untested data. The truth of the matter is that the amount of material available, even through regular channels, is so great that it may require the use of specialized information systems regardless of what may appear in the informal domain. Therefore, information systems should not be considered as competitive with the "long judicious procedure" of formal publication, but rather as complementary to it.

In any case, it is becoming increasingly likely that the user will have to avail himself of the services of the information specialist, who serves at the interface between an information system and those it is meant to serve.

THE INFORMATION SPECIALIST

Many different kinds of positions deal with the organization, storage, and retrieval of information. They may involve abstracting and indexing, systems analysis, computer programming, or possibly more traditional skills such as cataloging and classification. The setting or environment may range from a traditional library such as an academic or special library to a technical information center where books per se are not in evidence. The people who do the work are usually professional chemists with either formal or informal training in documentation or library and information science.

Perhaps the most important of these people, and certainly one of the most interesting, is the individual who must interact with the users and potential users of the information system or service. This person must know the subject field, how to use the system, and how to cope with the users. Adequate subject knowledge is usually assured if the person is a professional chemist, although he may need advanced study if there is substantial demand for information in specialized areas. Knowing how to use the system implies a knowledge of the nature of the collection or data base at hand, a good working familiarity with the interface language of the system, an understanding of the methods of file organization employed, and, if the system possesses a computer-based information retrieval capability, a knowledge of the proper methods of writing search logic to interrogate it. Coping with the users is probably more difficult.

The user may not have a clear perception of his own information need. Therefore, unless the query is of the "command" type, the information specialist or reference librarian would be wise to initiate the process of question negotiation in order to ascertain the exact nature of the user's problem. Naturally, this must be both an unobtrusive and inoffensive procedure. Helpful traits for the information specialist to cultivate are empathy, a good sense of analogy, and the ability to get to know the clientele. It is helpful, for example, to know precisely what the person's job is and whether he or she has any particular personality quirks. It is also useful to know what sort of answer will be considered acceptable; a certain amount of care must be exercised here, since some people may be satisfied with too little information. Because typical questions may be vague or even ill-informed, rephrasing them is an art that must be practiced. Tact and good psychology are extremely important traits; much can be learned, for example, from techniques such as the use of indirection, wherein the specialist may ask the user questions such as "If this is to be used for research on the synthesis of . . . , then . . ." Often the user may respond by saying that that was not at all what he had in mind but rather something else, which he will then go on to describe more adequately.

Several straightforward considerations can facilitate the use of an information service. First, there is that of the physical facilities themselves. The user of an information service is often out of his element, and he may not want to risk possible embarrassment by revealing his ignorance of the system, even when there is no reason for him to know about it. There is also evidence that users may avoid a library or information center in order to escape red tape. Consequently, it is advisable to make materials as accessible as possible and to avoid using "technical" library or information processing terminology insofar as possible. Another consideration is that of the user who wants to do his own search. Such a user may be sufficiently sophisticated in the use of the local facilities that he may wish to do things himself either to speed up the process or to avoid adding unnecessarily to the work of the information specialist. On the other hand, he may simply prefer the privacy of consulting information sources by himself. If this is the case, then the information specialist should guide the user to the best and most appropriate sources available, and, if possible, he should follow up later to see if he can be of further assistance.

Chemistry is blessed with a long history of concern for documenting its findings. It is also fortunate that there are some excellent guides to the traditional literature and information sources associated with the field. Those individuals wishing to do literature searches on their own can be referred to such fine works as those by Bottle,[5] Mellon,[6] or Crane, Patterson, and Marr.[7] Those interested in nontraditional information systems and services can be referred to the comprehensive and recent work by Kruzas.[8]

NOTES

1. Robert S. Taylor, "Question-Negotiation and Information Seeking in Libraries," *College & Research Libraries* 29 (1968): 182.

2. Ibid., p. 181.

3. Derek J. de Solla Price, *Little Science, Big Science* (New York: Columbia University Press, 1963), pp. 90-91.

4. William D. Garvey and Belver C. Griffith, "Scientific Communication: Its Role in the Conduct of Research and Creation of Knowledge," *American Psychologist* 26 (1971): 362.

5. R. T. Bottle, *The Use of Chemical Literature,* 2d ed. (London: Archon Books, 1969).

6. Melvin G. Mellon, *Chemical Publications, Their Nature and Use,* 4th ed. (New York: McGraw-Hill, 1965); and *Searching the Chemical Literature* (Washington, D.C.: American Chemical Society, 1964).

7. E. J. Crane, Austin M. Patterson, and Eleanor B. Marr, *A Guide to the Literature of Chemistry,* 2d ed. (New York: Wiley, 1957).

8. Anthony T. Kruzas, *Encyclopedia of Information Systems and Services* (Ann Arbor: Edwards Brothers, 1971).

3

Indexing Systems

Most of us have known at least one person who can find virtually anything on his desk, no matter how messy and cluttered that desk might be. On the spur of the moment, such individuals can reach into various improbable nooks and crannies, and retrieve whatever items are desired. For most of us, this random approach to the storage and retrieval of items can lead to disaster when we need to find something really important. In any case the efficacy of the technique is dependent on one individual's ability to recall what has been done with the item or items in question. Clearly, it would be more efficient to introduce order into such a situation, both to help the individual find things quickly and to help others who might have to use that desk.

Just as we can improve efficiency by introducing order to the items on a desk, we can increase the utility of an index or indexing system by providing an orderly and systematic arrangement of the terms to be used. In fact, a central purpose of indexing is to increase the efficiency of the information storage and retrieval process, and at the same time to make the system useful to an entire community of users having similar interests.

THE CONCEPT OF VOCABULARY CONTROL

Indexing may be thought of as a process of labeling items for future reference. Considerable order can be introduced to this process by standardizing the terms that are to be used as labels. This standardization is known as vocabulary control, the systematic selection of preferred terms. Consciously or unconsciously, we all exercise vocabulary control. The fact that this book is written in English rather than in French, Russian, or Swahili is evidence of control, standardization, and the introduction of order to the communication process. Similarly, when people converse they normally do so in a natural, mutually agreeable language. However, for indexing purposes, the problem of vocabulary control is not solved by choosing one particular natural language and sticking with it. Several linguistic problems have to be taken into consideration, the most important of which are (1) *synonymy,* (2) *semantic ambiguity,* and (3) the proper choice of *generic levels* of meaning.

Synonyms are two or more words having the same meaning. It should be fairly obvious that the use of synonyms in an index will lead to the scattering of information throughout the alphabet. For example, it is undesirable to use both 'acetaldehyde' and 'ethanal,' because the user of the index will then have to examine both places to make sure he has found everything that might be of interest. Similarly, terms and phrases denoting equipment, reactions, and so forth, should not be used if they are synonymous with other terms and phrases that appear elsewhere in the index.

Semantic ambiguity arises when a term or phrase has more than one meaning; this is very likely the most subtle of the three linguistic considerations mentioned. Terms or phrases that are written the same way but have different meanings are called *homographs,* and the condition is known as *homography.* Terms or phrases that sound alike but have different meanings are called *homonyms.* However, the term *homonymy* is used with reference to both homographs and homonyms—a clear-cut example of semantic ambiguity. Perhaps the best of the classic illustrations of semantic difficulties is given by the phrase "time flies like an arrow," which can be understood in three different ways: (1) time (the entity) passes by as quickly as an arrow in flight (i.e., *tempus fugit*); (2) time (maybe with a stopwatch) flies (the insects) as you would an arrow in flight; and (3) a certain species of flies, known as "time" flies, have developed a strange fondness for an arrow. The illustration may be amusing, but the problem is quite real. An example that is more germane to the field of chemistry would be the use of 'butenedioic acid,' which could refer either to fumaric acid or maleic acid, depending on the isomer.

Generic levels of meaning refer to the different levels of specificity which may be needed to adequately index an item. The term 'chemistry' might be very useful as an index term in a large research library, but its utility would be nil if all the items being indexed already dealt with some facet of chemistry. In like manner, 'organic chemistry' might be inappropriate if attention should be focused on the use of, say, the Friedel-Crafts reaction. The choice of the proper generic level will depend on the individual case, and it should be remembered that more than one term may be needed for adequate coverage.

Synonymy and semantic ambiguity are especially prevalent in English. The principal reason is that the English language is basically teutonic, but over the centuries has acquired so many words of Latin and Greek derivation that they now comprise almost 60 percent of the language. Naturally, this situation has resulted in considerable redundancy. An interesting account of this problem is given by Beckmann.[1]

SYNDETIC STRUCTURES

Clearly, both the indexer and the index user need a set of guidelines for the proper selection of terms. Syndetic structures provide these guidelines by showing the relationships among terms or concepts. These structures fall into two major categories: (1) classification schemes and (2) subject heading lists and thesauri.

Formal classification may be considered an orderly arrangement of knowledge serving to group like things or concepts. It is theoretically or philosophically based and is constructed in a predominantly deductive manner. Generally, a notation or coding is associated with a particular classification scheme and reflects its logical structure.

Unfortunately, the classification schemes with which most Americans are familiar, the Dewey Decimal Classification (DDC) and the Library of Congress Classification (LC), are not used for subject indexing. Rather, they are used for arriving at a single classification per item (usually a book) so that items which are superficially alike can be shelved together. This is not a bad practice since it can lead to fortuitous discoveries during browsing through the collection. However, because so many people associate classification with the shelving of books, too few investigations into the potential usefulness of classification for indexing purposes have been made in this country. Elsewhere, most notably in Great Britain and Europe, subject analysis and indexing by classification have been going on for a long time. The Universal Decimal Classification (UDC) and the so-called colon classification developed by the famous Indian classificationist Ranganathan have enjoyed the widest use. The fact that American usage of classification is restricted almost exclusively to shelving has led one expert in the field to refer to it as the "marking and parking" syndrome.

Classification schemes offer several advantages in unraveling the problems of vocabulary control. The problem of synonymy is usually handled by providing a "relative index" to the scheme. This index will include prominent synonyms and append the appropriate notation from the classification schedule to each of them. No matter how many different terms have the same referent, they will all be tied together by the use of only one classification number during indexing. Similarly, semantic ambiguity can be resolved by following each term in the relative index having more than one meaning with all the classification numbers which refer to the different concepts embodied in that single term. Since only the notation from the classification schedule will be used for indexing, there will be no ambiguity from the standpoint of storage or retrieval. Differing generic levels are usually dealt with by the construction of a hierarchical structure, which is normally reflected in the notation employed. Thus, in the UDC, pure science would be denoted by '5__,' chemistry by '54_,' and organic chemistry by '547.' The use of a decimal notation is, of course, purely arbitrary.

It is evident that the use of classification schemes alleviates many of the problems inherent in the use of terms taken from natural language.

However, classification schemes can create problems of their own. Because of their deductive and philosophical nature, they are often limited by their original structure and by the scope of man's knowledge at the time they were conceived. The following example, cited by Charles Meadow,[2] is an excellent case in point and shows what has happened in the Dewey Decimal classification:

6__	Technology (applied science)
62_	Engineering
629	Other branches of engineering
629.13	Aeronautics
629.138	Uses of aircraft
629.138 8	Space flight (inertial navigation)

There is obviously something contradictory in subsuming 'space flight' under 'uses of aircraft' and 'aeronautics,' but there is no choice in the matter. Problems of this sort are common in older *enumerative* classifications, which rely on listing all appropriate subjects. More modern classification schemes are generally of the *synthetic* type, in that they allow the user considerable freedom in coordinating and concatenating various elements of the classification in order to arrive at an appropriate label.

Although they will vary somewhat depending on the application, the criteria for evaluating a classification scheme (and the considerations for designing one) can be summed up as follows:

(1) Inclusiveness. Is the scope of coverage appropriate to the subject matter or field?

(2) Receptiveness to new subjects. Can new subjects be introduced readily, either by adding new elements or by coordinating elements already present?

(3) Logical division and subdivision of classes.

(4) Practicality of the notation. Is the notation reasonably brief? Does it adequately reflect the processes of division and subdivision? Are there mnemonics in the form of repeating symbols?

(5) Relative index. Is there one? If so, is it sufficiently comprehensive to handle most problems of synonymy and semantic ambiguity?

Subject heading lists and thesauri are similar in that both usually consist of alphabetically arranged terms, cross references, and notes to be used in indexing or searching a corpus of documents. Some specialists in documentation use the terms 'authority list,' 'subject heading list,' and 'thesaurus' interchangeably because of their evident similarity of design and purpose. However, to do so runs roughshod over distinctions which are not trivial. Traditional subject authority lists such as the

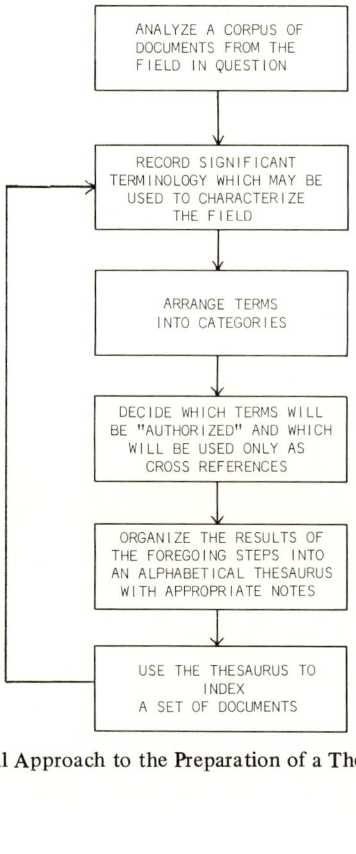

Figure 2–General Approach to the Preparation of a Thesaurus

Sears and Library of Congress subject heading lists are meant to deal with all of man's knowledge. Most, if not all, modern thesauri treat only a subset of this knowledge, namely a particular discipline or field of study. Moreover, the preparation of thesauri generally involves the sorting out of synonyms and homographs that actually appear in the text of the documents being indexed, rather than trying to anticipate the concepts that will be encountered. In short, traditional subject heading lists may have a philosophical basis rivaling that of formal classification, whereas thesauri are usually developed in a primarily inductive and pragmatic manner. This distinction may be made clearer by reference to Figure 2, which illustrates a general approach to the preparation of a thesaurus.

The choice of form for the main subject headings is of considerable importance because, among other things, it will determine the place in the alphabetical listing where items will be entered. There are five principal types of main heading form; they are shown, along with examples, in Figure 3.

Types	Examples
(1) Noun form.	Codeine. Explosives. Platinum.
(2) Noun preceded by adjective.	Laminated products. Molecular structure.
(3) Inverted form.	Acetic acid, trichloro- Light, infrared. Tin, analysis.
(4) Compound heading.	Plastics and Resinous products. Poisoning and Intoxication. Science and Technology.
(5) Phrase heading	Engineering as a profession. Lecture demonstrations in chemistry.

Figure 3–Principal Forms of Main Index Heading

Good syndetic structures must show the relationships among terms. In particular, they must take care of the potential problems caused by synonymy, semantic ambiguity, and differing levels of specificity. To do this, subject heading lists and thesauri employ three basic devices:

"See" references for synonymy; "See also" references for differing levels of specificity and certain unavoidable "near-" or "quasi-" synonyms; and parenthetical notes to resolve problems of semantic ambiguity or to provide general clarification.

"See" references are straightforward and indicate that one particular term is to be used in preference to another:

Ethanal. *See Acetaldehyde*
or
Ethanal. *Use Acetaldehyde*

The use of "See" in the cross reference is typical of more traditional subject heading lists, while "Use" is to be found in many thesauri. They are equivalent.

"See also" references are used to call attention to additional related terms which may be useful in indexing or searching. In most subject heading lists and library catalogs, the "See also" reference is used to refer to terms that are either more specific or are at the same generic level. This practice is based on the phenomenon, well known among reference librarians and other information specialists, that people ask questions at a more general level than is needed to satisfy their requests for information. It is quite reasonable to assume that when people do their own searching they will find those references most helpful which lead them to more specific information. However, there is no reason to believe that "See also" references to terms at higher generic levels would not occasionally be useful, and so this usage is to be found in many thesauri. The symbols 'RT,' 'NT,' and 'BT,' which are often used in place of "See also," stand for 'related term,' 'narrower term,' and 'broader term,' respectively.

The use of the "See also" type of reference to terms at different levels of specificity is fairly clear, but the concept of providing for near synonyms may require a little elaboration. Closely related terms may have to be used because of differences, sometimes subtle and sometimes not, in denotation or connotation. For example, making multiple entries for keto-enol tautomers is mandatory even though different names are being used for the same chemical compound! It is doubtful that many people would mind the relatively minor inconvenience of having "See also" references between 'Acetaldehyde' and 'Vinyl alcohol' because of the advantages of maintaining separate headings for items that conceptually are so different.

In order to make the relationships among terms as clear as possible, most subject heading lists and thesauri indicate the existence of cross references at the site of the preferred index term. "See" references are usually indicated by 'x' and "See also" references by 'xx'; the two symbols are read as 'see from' and 'see also from,' respectively. Some lists acknowledge the existence of cross references by listing all major synonyms in parentheses following the preferred index term.

Acetaldehyde
  x Ethanal
  xx Vinyl alcohol
Isoquinoline (2-*benzazine, leucoline*)

Thesauri employing 'RT,' 'NT,' and 'BT' types of relational indicators may also use corresponding symbols to denote the concepts of 'see also from, related,' 'see also from, specific,' and so forth.

Notes are usually displayed parenthetically following an entry. They are used to resolve ambiguities and to clarify the scope or inclusiveness of the terms associated with them. The following examples, which are from the *Chemical Abstracts* Master File of Cross References, illustrate these two principal uses for notes:

Mercury (*the planet*). *See Planets*

Power (*For consumption of or requirements for power see operations or procedures where power is used.*)

Figures 4 and 5 show sample entries from well-known and fairly typical subject heading lists and thesauri, respectively.

The criteria for evaluating and designing subject heading lists and thesauri may be summed up as follows:

(1) Terminology used. Is it appropriate for the field? Up-to-date? Accurate?

(2) Scope. Is it too broad or too narrow to cover the field adequately?

(3) Subdivisions. Are there reasonable subdivisions? (Sometimes it might make sense to subdivide headings by geography, other times chronologically, for example.)

(4) Definitions and notes. Are enough included for clarity?

(5) References. Are they adequate in both number and form?

(6) Format. Is it legible? (Look for such things as indention, type font, and the use of bold face and light face type.)

(7) Classification numbers. Is the listing keyed to any kind of classification scheme if appropriate?

Not surprisingly, there is some similarity between these criteria and those for evaluating classification schemes. One criterion, equally appli-

AUTOMATIC DATA PROCESSING (L)
XU PATTERN RECOGNITION (L)

AUTOMATION (J)
see also related
AUTOANALYSIS (E5)

AUTOMATISM (F1, F2)

AUTOMOBILE DRIVER EXAMINATION (I)

AUTOMOBILE DRIVING (I)
see also related
ACCIDENTS, TRAFFIC (G3)

AUTOMOBILE EXHAUST (G3)

AUTOMOBILES (J)

AUTONOMIC DRUGS (D5)
XU GANGLIONIC-STIMULATING AGENTS (D5)
XU NEUROPHARMACOLOGY (D5)

AUTONOMIC DYSFUNCTION (C10)
see also related
THERMOGRAPHY (E1)

AVIATION MEDICINE (G2)
see also related
ACCIDENTS, AVIATION (G3)
ALTITUDE (G3)
DECOMPRESSION SICKNESS (C14)
HUMAN ENGINEERING (F1, J)
XU AVIATION NURSING (G2)

AVIATION NURSING see under AVIATION MEDICINE (G2)

AVIDIN see under OVALBUMIN (D10)

AVITAMINOSIS (C13)
X VITAMIN DEFICIENCY (C13)

AVOIDANCE LEARNING (F1)
see also related
ESCAPE REACTION (F1)

AWARDS AND PRIZES (K)

AWARENESS see under COGNITION (F1)

AXILLA (A1)

AXILLARY ARTERY (A7)

Figure 4—Portion of the Medical Subject Headings (MESH) List Employed at the National Library of Medicine in the Production of *Index Medicus*

RT Formal organization
Organization theory
Information 1407
UF Recorded information
NT Bathythermograph data
—Data
Experimental data
Meteorological data
—Oceanographic data
Statistical data
Steam tables (thermodynamics)
—Tables (data)
Telemetering data
RT—Information theory
Information analysis centers
USE Information centers
Information capacity 0904
UF Bit capacity
BT Information theory
RT Entropy
Information centers 0502
A service directed toward collecting technical information in a specific area of effort and its evaluation and filtering into the form of condensed data, summaries, or state-of-the-art reports
UF Information analysis centers
Technical information centers
Information engineers
USE Documentalists
Information processing language
USE IPL (programming language)
Information retrieval 0502
UF Document retrieval
Records retrieval
RT Data retrieval
Descriptive cataloging
Documentation
Indexes (documentation)
Information retrieval effectiveness
Information systems
—Subject indexing
Information retrieval effectiveness 0502
UF Information retrieval precision
Recall ratio
Relevance ratio
RT Index term links
Information retrieval
Role indicators
Search structuring
—Subject indexing
Information retrieval precision
USE Information retrieval effectiveness
Automata theory
—Combinatorial analysis
Communication theory
Cybernetics
—Data processing
Data transmission
—Decision theory
Ergodic processes
—Fourier analysis
—Information
Information systems
—Linguistics
—Operations research
—Probability theory
—Random processes
Redundancy
Stationary processes
—Stochastic processes
Systems engineering
—Telecommunication
Infralow frequencies 2014
300 to 3000 Hz
BT Frequencies
Radio frequencies
RT Audio frequencies
Infrared absorption
USE Electromagnetic absorption
and Infrared radiation
Infrared analysis 1402
RT Gas analysis
Gas chromatography
Infrared beacons 1707
BT Beacons
Infrared equipment
RT Metascopes
Infrared binoculars 1705
BT Binoculars
Infrared equipment
Optical equipment
RT Infrared searchlights
Infrared telescopes
Infrared bolometers 1402
BT Bolometers
Detectors
Infrared detectors
Infrared equipment
Infrared thermal detectors
Measuring instruments
NT Infrared thermistors
Infrared bombing 1507
BT Bombing
RT Night bombing
Radar bombing
Strategic bombing
Infrared brazing 1308
BT Brazing

Figure 5—Portion of the Department of Defense (DOD) Thesaurus of Engineering and Scientific Terms

cable to the major types of syndetic structure, is consideration of the authority issuing the vocabulary controlling device. Normally, professional societies prepare better products for their fields than individuals or dissident splinter groups because they make greater efforts to obtain a consensus among specialists regarding appropriate terminology. Products prepared by government agencies are also usually reliable.

During the indexing procedures, indexers consult the syndetic structures to make sure that potential entries conform to the instructions in the references and notes. In general, only those references and notes that are actually used by the indexers during a given indexing period are printed in the index. Superfluous references and notes from the syndetic structure are not printed because they would confuse the user by leading him to believe that nonexistent entries were in the index.

As used here, the term 'syndetic structure' denotes subject heading lists, thesauri, and those classification schedules which are used for subject indexing. Subject analysis employing classification schemes, although uncommon in this country, is important for both theoretical and practical reasons. Interested readers might wish to examine the contents of the publication *Meteorological and Geoastrophysical Abstracts,* which uses the Universal Decimal Classification, and the publications by Schneider, which describe the system in use at the National Cancer Institute.[3]

INDEXING WITH LITTLE OR NO VOCABULARY CONTROL

Until now, we have been discussing rather careful indexing and the importance of vocabulary control. We have stressed the design and use of syndetic structures as a means of introducing order and maintaining consistency. While these are worthy techniques, they are rather time-consuming, and there are very definitely situations in which the currency of the information is more important than indexing consistency. When these situations arise, they are dealt with by minimizing or entirely eliminating all but the most important information processing steps prior to the actual printing of the index.

When such measures are taken, greater emphasis must invariably be placed on the search procedures, if thoroughness is still important, because the desired information may appear in widely scattered places and in many different contexts. One means of coping with the increased difficulties of searching is to employ various manual devices or an electronic digital computer in conjunction with post-coordinate indexing, a technique involving the coordination of terms after the fact instead of before as in the case of indexing from syndetic structures. This important concept of coordination is discussed more fully in Chapter 5.

The most popular and successful technique using little or no vocabulary control is keyword indexing by computer. Typically, the data (usually titles) are put into machine-readable form and input to a computer that has been programmed in an appropriate manner. The program may identify a word as a string of characters between two blanks and then compare each of these words with an internally stored list of "stop" words. "Stop" words are those which can be presumed to have no index value, such as articles, prepositions, and conjunctions. All words matching a "stop" word are rejected, while all others are retained. If the index is to follow the KWIC (*K*ey*W*ord-*I*n-*C*ontext) format, then sorting and spacing are performed to obtain a result such as that shown in Figure 6. If the KWOC (*K*ey*W*ord-*O*ut-of-*C*ontext) format is chosen, then the titles (or sentences) are "permuted" on the keywords as shown in Figure 7.

A modicum of vocabulary control can be introduced as part of the automated procedures involved in the preparation of keyword indexes. Lists of "unacceptable" terms can be stored in the computer and incoming words can be matched against them. If no match occurs, then the word is deemed "acceptable" and is printed out with the information associated with it; otherwise, the word is converted to a preferred index term prior to being printed. The difficulty here is that, while it is a relatively straightforward task to program the computer to recognize and convert synonyms, programming the machine to allow for homographs is almost impossible. Programs have been developed for syntactic analysis and for establishing the meaning of a word through its context, but these programs are never infallible and are always expensive. Therefore, the only practicable approaches to automated vocabulary control are probably those that treat data bases in which synonymy is high,

```
                                              SOME ALGORITHMS FOR ANALYSING ORIENTED GRAPHS.                              R0536
                                               TWO ALGORITHMS FOR BIPARTITE GRAPHS.                                       R0234
ISM AND AUTOMORPHISM GROUPS OF PLANAR /  ALTERNATIVE ALGORITHMS FOR DETERMINING ISOMORPHISMS AND AUTOMORPH GB248
R AN UNDIRECTED LINEAR GRAPH.                            ALGORITHMS FOR FINDING A FUNDAMENTAL SET OF CYCLES FO GB838
GRAPHS.                                                  ALGORITHMS FOR GENERALIZED STABILITY NUMBERS OF TREE  GB329
                                                      ON ALGORITHMS FOR PROBLEMS OF PATHS IN GRAPHS.               GB648
                                          REVISED MATRIX ALGORITHMS FOR SHORTEST PATHS.                            GB559
HAMILTON.                                                ALGORITHMS FOR THE PROGRESSIVE ORIENTATION OF GRAPHS. R0151
DIATE APPLICATIONS.                                CYCLE ALGORITHMS FOR UNDIRECTED LINEAR GRAPHS AND SOME IMME GB824
K.CIRCUIT-THEORY.                                        ALGORITHMS TO FIND THE MOST RELIABLE PATH IN A NETWOR R0912
RICES.                                                   ALGORITHMS TO REALIZE DIRECTED COMMUNICATION NETS.MAT GB593
                         AN APPRAISAL OF SOME SHORTEST PATH ALGORITHMS.                                            GB680
               THEORETICAL FOUNDATIONS FOR NEIGHBOR-LISTING ALGORITHMS.                                            GB685
                                        NONDETERMINISTIC ALGORITHMS.CYCLES.                                        GB257
                          PARALLELIZATION OF COMPUTATIONAL ALGORITHMS.DIRECTED.PARTITIONS.                         GB789
OF STRUCTURAL ISOMERS OF THE MORE IMPORTANT TYPES OF ALIPHATIC COMPOUNDS.CHEMISTRY.TREES.ENUMERATION.             R0103
ORING THE VERTICES OF A GR/  REDUCTION OF THE MEMORY ALLOCATION PROBLEM IN PROGRAMMING TO A PROBLEM IN COL CE059
                                                         ALMOST ALL TOURNAMENTS ARE IRREDUCIBLE.                   CM088
                          SEARCHING FOR THE SHORTEST PATH ALONG A GRAPH DURING VARIATIONS IN IT.                  GB791
OIS FIELDS.      ON A PROPERTY OF THE NORMED SPACE C-ALPHA AND ITS APPLICATIONS TO UNIMODULAR MATRICES.GAL R0132
                       THE NUMBER OF TREES WITH NODES OF ALTERNATE PARITY.ENUMERATION.COLORING.                   R0779
TE GRAPHS.                                               ALTERNATING HAMILTONIAN CIRCUIT IN TWO-COLORED COMPLE GB728
      ON NON-PRIMITIVE REGULAR GRAPHS OF DEGREE THREE.ALTERNATING PATHS.                                          R0572
ND AUTOMORPHISM AND AUTOMORPHISM GROUPS OF PLANAR /  ALTERNATIVE ALGORITHMS FOR DETERMINING ISOMORPHISMS A GB248
APH THEORY AND PARAMETRIC REPRESENTATIONS OF FEYNMAN AMPLITUDES.PHYSICS.                                      GR GB135
ON FOR GRAPHS.HOMOTOPY.                               AN ANALOGON OF THE FIXED-POINT THEOREM AND ITS APPLICATI GB057
                                   SOME ALGORITHMS FOR ANALYSING ORIENTED GRAPHS.                                  R0536
                                    SIGNAL FLOW GRAPH ANALYSIS AND FEEDBACK THEORY.LINEAR EQUATIONS.              R0440
                                          TOPOLOGICAL ANALYSIS AND SYNTHESIS OF COMMUNICATION NETWORKS.           R0489
OOK.                                          NETWORK ANALYSIS FOR PLANNING AND SCHEDULING.TRANSPORTATION.B R0047
                                            ON MATRIX ANALYSIS OF CONNECTIONS IN PARTIALLY DIRECTED GRAPHS.    CP049
ASTIC GRAPHS.                                            ANALYSIS OF DISCRETE MARKOV SYSTEMS BY MEANS OF STOCH GB807
                    THE APPLICATION OF GRAPHS TO THE ANALYSIS OF DISTRIBUTION OF LOOPS IN A PROGRAM.             GB784
                                                 MATRIX ANALYSIS OF ELECTRICAL NETWORKS.BOOK.                     R0532
              APPLICATIONS OF BOOLEAN MATRICES TO THE ANALYSIS OF FLOW DIAGRAMS.                                  R0686
```

Figure 6—Segment of a Typical KWIC Index

BETA-DIKETONES
MECHANISM OF THE REACTION OF DIAZOMETHANE WITH 2,4-PENTANEDIONE AND SYNTHESIS OF STERICALLY-HINDERED BETA-DIKETONES — WILLIAMS, ROBERT MAXWELL **ISU** 19/08 /1924 **58-7570**
SYNTHESIS AND PROPERTIES OF BIS(BETA-DIKETONES) — MARTIN, DEAN FREDERICK **PSU** 19/11 /2768 **59-767**
SOME ACYLATIONS AND ALDOL TYPE CONDENSATIONS INVOLVING DICARBANIONS OF BETA-DIKETONES. CERTAIN CYCLIZATIONS. — LIGHT, ROBLEY JASPER **DUKE** 21/12 /3630 **61-1444**
ACYLATIONS AND OTHER CONDENSATIONS OF DIANIONS OF BETA-DIKETONES, IMIDES AND AMIDES. ACETYLATION OF 1,3,5-TRIKETONES BY BORON FLUORIDE. — WORK, STEWART DEAN **DUKE** 24/03 /983 **63-6619**
CLEAVAGE OF BETA-DIKETONES AND RELATIVE REACTIVITIES OF ORTHO AND PARA ALKYLBENZYL CHLORIDES — RIHAN, TALAAT IBRAHIM EL SAYED **U/TX** 25/07 /3847 **64-6628**
ALKYLATIONS AND OTHER CONDENSATIONS OF DI- AND TETRAALKALI SALTS OF BETA-DIKETONES — HAMPTON, KENNETH GERALD **DUKE** 02/601/90 **65-7272**
I. ATTEMPTED SYNTHESES OF BRIDGE-HEAD HYDROPHENANTHRENES II. REACTION OF GRIGNARD REAGENTS WITH BETA-DIKETONES — HARRISON, ERNEST AUGUSTUS JR. **U/MD** 27/04B/1088 **66-9291**
KINETICS AND MECHANISMS OF THE ACID CATALYZED CLEAVAGE OF BETA-DIKETONES — EL-EMARY, MOHAMOUD MOHAMED **U/TX** 28/06B/2339 **67-14,905**
CONDENSATIONS INVOLVING DIANIONS OF BETA-DIKETONES AND PHENYLACETAMIDES. STEREOCHEMICAL STUDIES — SCHRILTZ, DON MORRIS VON **DUKE** 28/02B/0599 **67-9759**

BETA-DIMETHYLGLYCI
THE PREPARATION AND REACTIONS OF ALPHA-PHENYL-BETA,BETA-DIMETHYLGLYCIDONITRILE AND THE CORRESPONDING AMIDE. — YOUNG, RAYMOND HINCHCLIFFE JR. **U/ME** 22/06 /1834 **61-5211**

BETA-DINITROALKANE
PREPARATION OF 1-ALKOXY-2-NITROALKANES AND THEIR CLEAVAGE TO NITROALCOHOLS. REARRANGEMENT OF SECONDARY BETA-DINITROALKANES TO 3,5-DIALKYLISOXAZOLES. — MARKOFSKY, SHELDON BRUCE **PURDUE** 23/10 /3636 **63-2111**

BETA-DIOLS
THE CONDENSATION OF HYDRAZINES AND BETA-DIOLS WITH 2,5-HEXANEDIONE AND 2,5-DIETHOXYTETRAHYDROFURAN. — SHEELEY, RICHARD M. **BRIGY** 25/03 /1583 **64-8967**

BETA-DISUBSTITUTED
REACTIONS OF BETA-DISUBSTITUTED ENAMINES WITH ISOCYANATES AND ISOTHIOCYANATES. — MINA, GEORGE LOUIS **STEVENS** 27/11B/3862 **66-12795**

Figure 7—Example of a KWOC Index

semantic ambiguity is minimal, and terms are mostly nouns with physical objects as their referents. Lists and tables of chemical compounds are good examples of this kind of data base.[4,5]

Since most keyword indexes are prepared without vocabulary control, they are not usually cumulated. After several indexes have been produced, the entries are not merged to produce a larger volume because this would exacerbate the problem of information scattering.

SPECIAL TYPES OF INDEXES

Thus far only subject indexing and word indexing have been discussed in this chapter, the latter differing from the former in that terms are taken as they come with no attempt being made to classify the concepts underlying them. In addition, there are other important types of indexes such as citation indexes, formula indexes, and ring indexes.

Citation indexing is a technique based on the presumption that a reader interested in one particular document will also be interested in all the others which cite it. A citation index arranges the bibliographic information in such a way that the user can look up a given document, find the works which cite it, if any, and then branch forward in time until the supply of relevant newer documents is exhausted. A portion of a citation index is shown in Figure 8.

While citation indexing is a very old technique, until recent years its chief success had been in the field of law because of the importance of citing legal precedents. Following the appearance, several years ago, of the new publication *Science Citation Index,* citation indexing has also

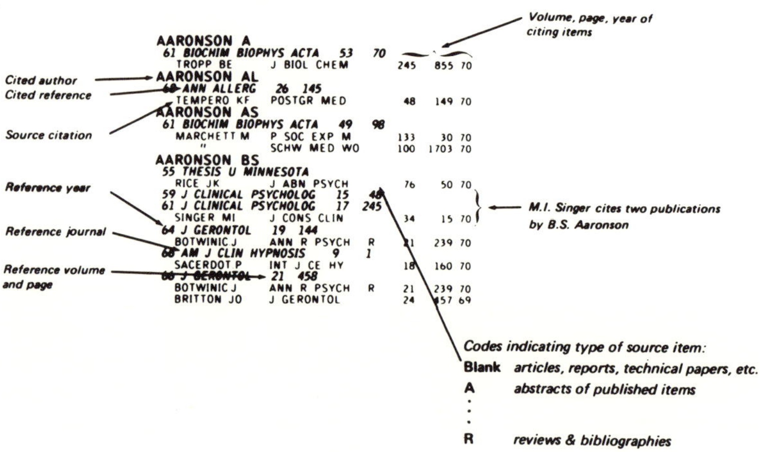

Figure 8—Annotated Portion of the Science Citation Index (published by the Institute for Scientific Information, Philadelphia)

proven valuable in chemistry and in other scientific and engineering disciplines.[6] Strong advocates of this technique are fond of pointing out that documents and articles which can not be found during a conventional subject index search can be uncovered by using a citation index. This is true. However, so is the converse, and important articles may be overlooked if one uses a citation index to the exclusion of good subject indexes. Therefore, the researcher who has to do a thorough, scholarly retrospective search would be well advised to use both types of index since they are complementary rather than mutually exclusive.

Formula indexes have come into being as companions to chemical subject indexes because of the complexity of chemical nomenclature, which has all but reduced chemists to the status of laymen when it comes to searching the literature of their own field. Even if a chemist is unable to arrive at a satisfactory name for a compound, however, he can invariably calculate its empirical formula. After that, it is relatively simple to arrange the elemental symbols in the order used by the formula index and to check for a match. There are two major systems for ordering the chemical symbols: the Hill and the Richter systems. Order in the Hill system is alphabetical by elemental symbol unless carbon is present. When this is the case, the formula begins with C followed immediately by H if hydrogen is also present in the molecule. All other elemental symbols then follow alphabetically. The Richter system starts with C, H, O, N, Cl, Br, I, F, S, and P, and then follows with any others in alphabetical order. Naturally, an empirical formula may not be unique; therefore, two or more names often follow the formula in an index, indicating which isomers have appeared in the documents covered by the index (see Figure 9 for illustrations). Since it is much easier to visualize or draw a structure from a given name than it is to derive a name from a given structure, the average chemist usually has no difficulty using a formula index. Various problems of nomenclature and special considerations imposed by index structure are discussed in Chapter 7.

There is one type of formula indexing which is somewhat like KWIC and KWOC indexing. Compounds containing hetero atoms are indexed under their molecular formulas as many times as there are hetero atoms in the molecule, with the hetero atom prominently displayed in some

$C_{14}H_{28}$
Cyclohexane, 1,4-di-*tert*-butyl-
cis- [*4789-34-8*], 3342b
trans- [*4789-35-9*], 3342b
—, 1,2-diisobutyl- [*30313-64-5*], P 22443b
Cyclooctane, 1,4-diisopropyl- [*30313-65-6*], P 22443b
Cyclotetradecane [*295-17-0*], 22373d, P 141104b
3-Hexene, 3-*tert*-butyl-2,2,5,5-tetramethyl- [*28923-90-2*], 140746a
Tetradecene
dimer, P 87415p
1-Tetradecene [*1120-36-1*], 12547z, P 78103h, P 140924g
polymers [*25608-58-6*], P 14796d
polymers, isotactic [*26746-87-2*], 92194v, 92198z
polymer with 1-octadecene [*30660-52-7*], P 14796d
7-Tetradecene [*10374-74-0*], P 44064f, P 87311b, P 111530v, P 124833w
$C_{14}H_{28}AgNS_2$
Silver, (heptylhexyldithiocarbamato)- [*31333-99-0*], 80413c
$C_{14}H_{28}AlO_4$
Aluminate(1-), dihydrobis(2-methoxycyclohexanolato)-
sodium [*31532-26-0*], P 89240b
$C_{14}H_{28}As_4CoN_2S_2$
Cobalt(1+), bis(isothiocyanato)bis[vinylenebis[dimethylarsine]]-
hexafluorophosphate(1-), *trans*- [*31159-98-5*], 76487n
tetrakis(isothiocyanato)cobaltate(2-) (2:1), *trans*- [*31226-03-6*], 76487n
$C_{14}H_{28}As_4N_2NiS_2$
Nickel, bis(isothiocyanato)bis[vinylenebis[dimethylarsine]]-
trans- [*31223-31-1*], 76503q
$C_{14}H_{28}AuNS_2$
Gold, (heptylhexyldithiocarbamato)- [*31328-96-8*], 80413c
$C_{14}H_{28}BrNO$
Acetamide, 2-bromo-*N*-dodecyl- [*15537-87-8*], P 98504t
$C_{14}H_{28}CdN_2S_4$
Cadmium, (diisopropyldithiocarbamato)- [*31312-98-8*], P 4471e

(a)

$C_{14}H_7O_3Cl$ 4-Chlor-1-oxy-anthrachinon **8**, 340 (651).
6-Chlor-1-oxy-anthrachinon **8**, 340.
7-Chlor-1-oxy-anthrachinon **8**, 340.
1-Chlor-2-oxy-anthrachinon **8**, 344 (658).
3-Chlor-2-oxy-anthrachinon **8**, 344.
$C_{14}H_7O_3Cl_3$ 3.4.6- oder 3.5.6-Trichlor-2-benzoyl-benzoesäure **10**, 750.
3.6-Dichlor-2-[4-chlor-benzoyl]-benzoesäure **10** (357).
4(oder 5)-Chlor-2-[2.5-dichlor-benzoyl]-benzoesäure **10** (357).
$C_{14}H_7O_3Cl_5$ Pentachlorphenyl-benzyl-carbonat **6**, 437.
$C_{14}H_7O_3Br$ 4-Brom-1-oxy-anthrachinon **8**, 341 (652).
3-Brom-2-oxy-anthrachinon **8**, 344.
x-Brom-fluorenon-carbonsäure-(4) **10** (370).
$C_{14}H_7O_3Br_5$ x.x.x.x.x-Pentabrom-[3-acetyl-naphthyl-(1)-acetat] **8**, 151.
$C_{14}H_7O_3As$ Anthrachinonyl-(1)-arsenoxyd **16** (441).
$C_{14}H_7O_4N$ 1-Nitro-anthrachinon **7**, 791 (415).
2-Nitro-anthrachinon **7**, 792 (415).
2-Nitro-phenanthrenchinon **7**, 806 (419).
3-Nitro-phenanthrenchinon **7**, 806 (419).
4-Nitro-phenanthrenchinon **7**, 807 (419).
Bei 215—220° schmelzendes x-Nitro-phenanthrenchinon **7**, 807.
Bei 260—266° schmelzendes x-Nitro-phenanthrenchinon **7**, 807 (419).
$C_{14}H_7O_4N_3$ 2.6-Dioxy-4-[3.4-methylendioxy-phenyl]-3.5-dicyan-pyridin **27**, 532.

(b)

Figure 9—Examples of Formula Indexes Prepared According to (a) the Hill System and (b) the Richter System

fashion. Perhaps the best two examples of such indexes are the HAIC (*H*etero-*A*tom-*I*n-*C*ontext) Index, which supplements the Formula Index to *Chemical Abstracts*, and the Rotaform Index, which appears in

$C_{58}H_{80}N_6$	Ni	O_4
Co_2	Ni	O_4
CrFe	Ni	O_4
Cr_2	Ni	O_4
Fe_2	Ni	O_4
Ga_2	Ni	O_4
La_2	Ni	O_4
Mn_2	Ni	O_4
$(C_2H_8N_4$	Ni	$O_4)_x$
$(C_3H_{10}N_4$	Ni	$O_4)_x$
$(C_4H_{12}N_4$	Ni	$O_4)_x$
$(C_6H_2N_2$	Ni	$O_4)_x$
$(C_8H_{12}N_4$	Ni	$O_4)_x$
$(C_{10}H_{16}N_4$	Ni	$O_4)_x$
$(C_{12}H_8N_2$	Ni	$O_4)_x$
$(C_{14}H_{24}N_4$	Ni	$O_4)_x$
$(C_{18}H_{16}N_4$	Ni	$O_4)_x$
$(C_{18}H_{20}As_2$	Ni	$O_4)_x$
$(C_{22}H_{18}N_4$	Ni	$O_4)_x$
$(C_{22}H_{24}N_4$	Ni	$O_4)_x$
$(C_{24}H_{22}N_4$	Ni	$O_4)_x$
$(C_{38}H_{28}As_2$	Ni	$O_4)_x$
$(C_{42}H_{36}As_2$	Ni	$O_4)_x$
$C_{12}H_8$	Ni	$O_4 \cdot C_2H_8N_2 \cdot 2H$
$C_{26}H_{20}N_{10}$	Ni	$O_4 \cdot 2C_5H_5N$
$C_{12}H_{28}N_8$	Ni	$O_4 \cdot 2Cl$
$C_{27}H_{27}N_6$	Ni	$O_4 \cdot ClO_4$
$C_{29}H_{31}N_6$	Ni	$O_4 \cdot ClO_4$
$C_{72}H_{60}As_4$	Ni	$O_4 \cdot 2ClO_4$
$C_4H_8N_2$	Ni	$O_4 \cdot 2H_2O$
$C_8H_{13}N_5$	Ni	$O_4 \cdot {}^1/_2H_2O$
$C_{10}H_{14}$	Ni	$O_4 \cdot 2H_2O$
$C_{12}H_{16}N_6$	Ni	$O_4 \cdot H_2O$
$C_{12}H_{24}N_2$	Ni	$O_4 \cdot 2H_2O$
$C_{19}H_{32}Cl$	Ni	O_4P
$C_{28}H_{29}$	Ni	O_4P
$C_{24}H_{54}$	Ni	O_4P_2S
C_4H_{12}	Ni	$O_4P_2S_4$
$C_8H_{16}Cl_4$	Ni	$O_4P_2S_4$
C_8H_{20}	Ni	$O_4P_2S_4$
$C_{14}H_{20}N_2$	Ni	$O_4P_2S_4$
$C_{14}H_{22}N_2$	Ni	$O_4P_2S_4$
$C_{16}H_{20}N_2$	Ni	$O_4P_2S_4$
$C_{18}H_{24}Cl_4N_2$	Ni	$O_4F_2S_4$
$C_{18}H_{28}N_2$	Ni	$O_4P_2S_4$
$C_{18}H_{30}N_2$	Ni	$O_4P_2S_4$
$C_{20}H_{24}Cl_4N_2$	Ni	$O_4P_2S_4$
$C_{20}H_{28}N_2$	Ni	$O_4P_2S_4$

(a)

Mg *CONT.*

F6 I C8 H3
F7 I C9 H4
I C9 H9 N2 O
I C9 H11 O2
I C9 H13 N2 O
I C11 H9 O
I C15 H17 N2 O
I C17 H21 N2
I C22 H16 N
I C23 H38 N O

Mg2

C17 H31 N
Cl2 C20 H17 N
Cl2 C21 H19 N
Cl2 F8 C12 O

Mn

C28 H44

Mo

C10 H10
C20 H30
C20 H32
C37 H49 N4 O2
C37 H51 N4 O2
Cl2 C20 H30

Mo2

C20 H20
C40 H60

Na

As C28 H20
B C19 H42 O
Br C7 H4 N6 O2
Br C8 H6 N6 O2
Br C10 H8 N2 O2
Br C15 H11 N O
Br C18 H18 O
Br Cl S2 C15 H11 N2 O3
Br F3 C3 H3 O
Br S2 C15 H12 N2 O3
Br S2 C16 H14 N2 O3
C3 H4 N O2
C5 H6 N O2
C6 H3 N2 O2
C6 H7 N2 O2

(b)

Figure 10–Special Types of Formula Indexes: (a) the HAIC Index and (b) the Rotaform Index

Index Chemicus. Detailed descriptions of these two indexes appear in the introductions to each issue of *Chemical Abstracts* and *Index Chemicus.* Brief examples of them are shown in Figure 10.

Because of the importance, variety, and complexity of ring systems, special indexes based on their skeletal structure have been developed. The best known are *The Ring Index* by Patterson, Cappell, and Walker,[7] and the *Index of Ring Systems,* which supplements the subject indexes of *Chemical Abstracts.* Both employ a special code based on the cyclic structure of the compounds regardless of the presence of hydrogen atoms or substituents. The organization of these indexes is illustrated in Figure 11, which is an adaptation from an introduction to the *Index of Ring Systems.* First, the systems are arranged in increasing numerical order according to the total number of component rings. This number is calculated by determining how many scissions are needed to convert the entire ring structure to an open-chain system.

22,22
$C_{20}N_2$–$C_{20}N_2$
1,12-Diazabicyclo[10.10.10]dotriacontane
C_{22}–C_{22}
Bicyclo[10.10.10]dotriacontane

3-RING SYSTEMS

3,3,3
C_2O–C_2O–C_3
1,5-Dioxadispiro[2.0.2.1]heptane
C_3–C_3–C_3
Tricyclo[1.1.0.$0^{2,4}$]butane
3,3,4
C_2N–C_2N–C_4
1,5-Diazadispiro[2.0.2.2]octane
1,6-Diazadispiro[2.1.2.1]octane
C_3–C_3–C_4
Dispiro[2.0.2.2]octane
Dispiro[2.1.2.1]octane
Spiro[bicyclo[2.1.0]pentane-5,1′-=
cyclopropane]
Tricyclo[3.1.0.$0^{2,4}$]hexane
Tricyclo[2.1.0.$0^{2,5}$]pentane
P_3–P_3–P_4
Hexaphosphatricyclo[3.1.0.$0^{2,4}$]hexane
3,3,5
C_2O–C_2O–C_4O
3,5,7-Trioxatricyclo[4.1.0.$0^{2,4}$]heptane
C_2O–C_2O–C_5
3,7-Dioxatricyclo[4.1.0.$0^{2,4}$]heptane
C_2O–C_3–C_5
3-Oxatricyclo[4.1.0.$0^{2,4}$]heptane
C_3–C_3–C_4N
8-Azadispiro[2.0.2.3]nonane
5-Azatricyclo[4.1.0.$0^{2,4}$]heptane
C_3–C_3–C_4O
5-Oxatricyclo[4.1.0.$0^{2,4}$]heptane
3-Oxatricyclo[3.1.0.$0^{2,6}$]hexane
C_3–C_3–C_5
Dispiro[2.0.2.3]nonane
Dispiro[2.1.2.2]nonane
Tricyclo[4.1.0.$0^{1,3}$]heptane
Tricyclo[4.1.0.$0^{2,4}$]heptane
Tricyclo[3.1.0.$0^{2,6}$]hexane

Figure 11–Index of Ring Systems Used by the Chemical Abstracts Service

Second, the ring systems are arranged in increasing numerical order with respect to the number of elements in each component ring. Third, the elemental analysis of each component ring is displayed using the Hill system. Finally, the names of the ring systems are arranged alphabetically under these elemental analyses as shown.

The Ring Index provides names for the various ring systems and also cites the earliest references to the systems in the chemical literature. The *Index of Ring Systems,* on the other hand, provides only preferred index terms, so that the searcher must look them up in the accompanying subject index to *Chemical Abstracts* to find further information about the compounds containing a given ring system.

CONCLUSION

The design and operation of indexing systems have been and remain a mixture of science and art. They may always remain so. Recently, however, there has been interest in examining the theoretical basis of indexing and the flip side of the coin—the retrieval process. After reading Chapter 5, those interested might wish to examine the work of Landry and Rush and the references they cite.[8]

NOTES

1. Peter Beckmann, "Computerization of English," *IEEE Spectrum* 8, no. 12 (1971): 20-27.
2. Charles T. Meadow, *The Analysis of Information Systems: A Programmer's Introduction to Information Retrieval* (New York: Wiley, 1967), p. 23.
3. John H. Schneider, "Selective Dissemination and Indexing of Scientific Information," *Science* 173 (1971): 300-308.
4. For a good discussion of automatic information storage and retrieval, see Gerard Salton, *Automatic Information Organization and Retrieval* (New York: McGraw-Hill, 1968).
5. Charles H. Davis, "An Approach to Automated Vocabulary Control in Indexes of Organic Compounds," *Journal of Chemical Documentation* 7 (1967): 131-134; "An Approach to Automated Vocabulary Control in Indexes of Organic Compounds, II," *Journal of Chemical Documentation* 9 (1969): 252-256.
6. Eugene E. Garfield, " 'Science Citation Index'—A New Dimension in Indexing," *Science* 144 (1964): 649-654.

7. Austin M. Patterson, Leonard T. Cappell, and Donald F. Walker, *The Ring Index,* 2d ed. (Washington, D.C.: American Chemical Society, 1960), plus supplements.

8. Bertrand C. Landry and James E. Rush, "Toward a Theory of Indexing," *Proceedings of the American Society for Information Science* 5 (1968): 59-64, and "Toward a Theory of Indexing–II," *Journal of the American Society for Information Science* 21 (1970): 358-367.

4

Abstracts and Abstracting

A discussion of the concepts "abstract" and "abstracting" is always fraught with difficulty because the data upon which one must draw cover a great time span and because, until very recently, the concepts have been part of a venerable, empirical science. Thus, while the term "abstract" implies some kind of reduction of a corpus of things, the nature of the reduction is poorly defined. Similarly, while the term "abstracting" implies a method of reduction, the known methods are also purely descriptive. In this chapter some of the history of abstractions, the significance of the term "abstract," and both manual and automated abstracting systems are discussed.

WHAT IS AN ABSTRACT?

Abstraction pervades the whole of science, and suggests a transition from specific to general, from individual observations to class description. Thus, molecular structural representation of chemical substances is an abstraction based on the detailed examination of many, but not by any means all, individual chemical species. A description of some research is an abstraction of the actual research.

Therefore an abstract might be described as the quintessence of that from which it derives, or as a document shorn of detail. This notion is most often expressed by saying that an abstract is a condensed or abbreviated version of a document.

When the term "abstract" is used alone, its meaning seems ambiguous. As such it may refer to an abstract property or expression. ("World War II" is a concrete expression, while the word "war" is abstract). It may refer to title to real property. (In this sense an abstract is a written, connected, chronological summary of the essential portions of all recorded documents and facts which can be discovered by a search of the public records of the jurisdiction within which the realty is located.)[1] Or it may refer to a summary of the principal findings of the work reported in a paper.[2] Only the last of these three senses of "abstract" is of importance here, although it should be emphasized that the three senses are really quite similar in intent, all of them striving to represent the *essential* qualities of the thing(s) abstracted.

HISTORICAL DEVELOPMENT OF ABSTRACTING

The Royal Society of London was founded in 1662 to foster scientific inquiry and provide for the publication of research results. The proceedings of its meetings, the Philosophical Transactions, were first published in 1665, but it was not until the nineteenth century—almost 200 years after the first serial publication appeared—that the data were collected and summarized on a regular basis. In 1821, Berzelius began his *Jahresberichte über die Fortschritte der physischen Wissenschaften.* Berzelius' effort was preceded by a pharmaceutical "yearbook," *Berlinisches Jahrbuch für die Pharmacie und die damit verbunden Wissenschaften,* which was started in 1795 with a similar purpose. These publications were yearbooks only in the sense that they were annual summaries rather than more frequently appearing periodicals. As the number of articles increased and the need for more frequently published summaries or abstracts was felt, *Pharmaceutisches Zentralblatt* was begun in 1830. Subsequently, the trend toward the greater number, frequency, and specialization of abstracting publications has resulted in the concurrent appearance of a large proportion of them geared to the research needs of modern chemists. It is fashionable to illustrate the parallel growth of the primary journal literature and the secondary (abstract) journal literature by means of a graph such as that shown in Figure 12. In any event, the importance of abstracts as a means of documenting scientific research seems amply demonstrated.

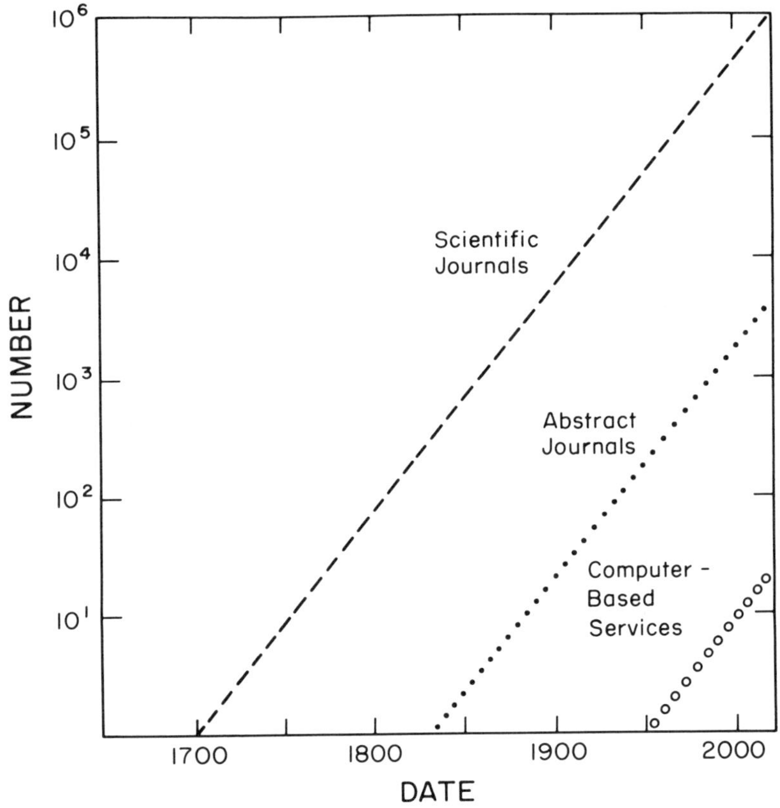

Figure 12–Comparison of the Growth of the Primary Literature (—), the Secondary Literature (...), and Computer-based Services (estimated) [ooo]

Traditional Views of "Abstract"

The American Chemical Society's handbook for authors admonishes authors to be aware of the "importance taken on by his abstract."[3] It defines the "ideal abstract" as one which

> will state briefly the problem, or purpose of the research when that information is not adequately contained in the title, indicate the theoretical or experimental plan used, accurately summarize the principal findings, and point out major conclusions. The author should keep in mind the purpose of the abstract, which is to allow the reader to determine what kind of information is in a given paper and to point out key features for use in indexing and eventual retrieval. It is never intended that the abstract substitute for the original article, but it must contain sufficient information to allow a reader to ascertain his interest. The abstract should provide adequate data for the generation of index entries concerning (both) the kind of information present and (the) key compounds (reported).

Chemical Abstracts Service's *Directions for Abstractors* states that "*CA* publishes informative abstracts which contain the significant content of published works." "A *CA* informative abstract is a concise rendition of the significant content of a bibliographically cited paper or report which provides enough of the new information contained in the work with sufficient abbreviated details to enable a reader to determine if it is necessary to consult the complete work." This description of an "informative abstract" leaves a great deal to the discretion of the abstractor, but the publication gives additional guidance on significant content: "The following components are considered to be significant in the contents of an article and are included in *CA* informative abstracts:

> The purpose and scope of the work, if it is not evident from the title.
> New reactions, compounds, materials, techniques, procedures, apparatus, data, concepts, and theories.
> New applications of established knowledge.
> The results of the investigation.
> The author's interpretation of the results and his conclusions derived from them."[4]

Similar descriptions of abstracts may be found in the publications of other primary and secondary information services.

Wyllys provides a more succinct, if not more precise, description of the term "abstract."[5] He states that an abstract is:

a) a description of, or restatement of, the essential content of a document
b) which is phrased in complete sentences (except for bibliographic data)
c) and which usually has a length in the range from 50 to 500 words.

The three descriptions of abstracts given above all lack definitional precision, but they do provide some points on which a definition can be formulated. Thus, the abstract should consist of complete sentences rather than keywords, phrases, or association maps.[6] The abstract should usually be short, although the relation between the length of the article and its abstract is prescribed in many different ways.[7] The essential content of the article, as reflected in the purpose of the work, results and conclusions, new data, and the like, should be included in the abstract. The difficulty with this last statement lies in determining what is the "essential content"; this is a problem central to all of language processing.

Types of Abstract

There are many types of abstract. Among these are:

- informative abstracts
- indicative abstracts
- extracts
- telegraphic abstracts

The first two types emphasize the conceptual content of the original document, while the latter two are more concerned with the structure of the original document.

Informative abstracts contain some or all of the information conveyed by the original document. A typical sentence in informative abstracts is: Propenoyl chloride (from 173 g. propenoic acid and 500 g.

PCl_5 in 300 g. dry benzene with stirring and cooling) added to 350 g. aluminum chloride in 1000 g. benzene at 65–70°, stirred 2 hours, and poured onto ice, the crude product distilled or recrystallized from methanol yielded 230 grams of 1,3-diphenyl-1-propanone, m.p. 72°.

Indicative abstracts merely state that certain information is contained in the original document. As rephrased in the indicative abstract, the sentence given above would read: the preparation of propenoyl chloride and its reactions with benzene and its derivatives are described. Examples of both informative and indicative abstracts are given in Figure 13.

a. ABC is judged a more satisfactory process for the waterproofing of synthetic textiles than XYZ. The ABC process yields a product of 20 percent greater durability as judged by standard test #1234. It also yields a better appearing product based on the votes of a panel of 25 textile finishing specialists. The cost of ABC is claimed to be 10 per cent less per square yard although specific cost data are not given. The two processes are about equal in processing speed.

b. The finishing of textiles by process ABC to achieve water repellency is considered superior to finishing by process XYZ. Factors considered include durability, appearance, cost, and speed.

Figure 13–Examples of (a) Informative and (b) Indicative Abstracts

Extracts may be either informative or indicative. The term extract signifies that some part of the original document has been carried over, without alteration, into the abstract. For example, the sentence: "In this paper we describe the synthesis of propenoyl chloride and its reactions with benzene and certain of its derivatives" might be found in an original document and in its abstract as well. If the abstract consists entirely of sentences from the original document, it is called an extract.

A *telegraphic abstract* consists of keywords or phrases that are usually taken from the original document. Few syntactic relations are preserved in such an abstract; hence, there is no sentence structure in a telegraphic abstract. A portion of a telegraphic abstract might have the appearance: propenoyl chloride; preparation; reaction; benzene; benzene derivatives.

Many other types of abstract have been described, but the four mentioned above are the most important. All others may be considered to be variants or combinations of these types.

Definition of "Abstract"

An abstract can be viewed in two ways: as a structural element to be formed and manipulated, or as an intensional element. Most, if not all, current attempts to define the term "abstract" mix these two views and consequently fail to provide a definition.

The following semiformal definition of an abstract, based on structural considerations, is offered to provide a workable and useful basis for a better understanding of the abstracting process. (A more complete formal definition may be found in the works cited in footnotes 8 and 9.)

An abstract, *A*, is a set of sentences, *s*, such that

$$A = \{s | s \in D\}$$

where *D* is the document abstracted, and such that certain transformations on the *s* are allowed:

concatenation
truncation
phrase deletion
voice transformation
paraphrase
division
word deletion

What does this definition do for us? First, it provides a more precise means, than those described earlier, of determining whether a document is an abstract. Thus, one of its most important advantages is that it distinguishes abstracts from critical reviews or other literary inventions of the abstractor. Second, it makes no mention of "content" or "meaning," notions which are difficult to deal with and which may vary with the *purpose* of an abstract (which will be dealt with in the following section). Hence, the definition distinguishes between what an abstract *is* and what an abstract *does*. Third, the definition allows for a certain stylistic freedom, but stylistic freedom does not encompass editorial comment, which is clearly precluded by the definition. Finally, the definition is applicable to both human and nonhuman abstractors, and so permits comparison of the abstracts produced by different abstracting systems.

The definition given above does not prescribe the length of an abstract. Length is a function of the *purpose* of an abstract and therefore is not of the essence of the concept "abstract." The length of an abstract is accordingly dictated by the abstracting system rather than by the definition of the product of the system.

Intensional Properties of an Abstract

While the definition provided in the preceding section is useful, it is incomplete because it does not touch on the intension of an abstract. Abstracts are usually placed into one or the other of two intensional classes: *informative* or *indicative*. While many variations on these two classes of abstract have been described,[10] in practice, most abstracts prove to be both informative and indicative. Therefore, it is perhaps less important to consider abstracts as belonging to one or the other of these classes than it is to consider the user population they are to serve.

To know the needs and desires of a system's user population is no easy accomplishment. Assuming that existing abstracting services have this knowledge, it can be observed that *Chemical Abstracts*, for example, gives greater emphasis to current work in chemistry than to articles discussing historical data (reviews, etc.). The reason for this observation is that articles of the first type are represented by informative abstracts, while the historical type are represented by indicative abstracts. This leads to the conclusion that the users of *Chemical Abstracts* are, on the whole, more concerned with current work than with historical data. More important, for this discussion, is the fact that the purpose of the abstract dictates what particular set of sentences will go into it. Therefore, the structure and the intension of the abstract are *not* independent, and so the definition given above must be modified to account for this relationship between structure and intension. Such a modified definition is described elsewhere.[11]

ABSTRACTING

Thus far, the concept of an abstract has been treated at some length, and attention has been focused on the questions:

(1) What is an abstract?
(2) What is the purpose of an abstract?
(3) How may an abstract be produced?

In this section, the last question is considered in greater detail. In general, an abstract is produced by an abstractor or an *abstracting system.* The term abstractor seems to imply a human agent, but the term abstracting system can be applied with equal ease either to humans or to machines. The human abstracting systems will be considered first, followed by a discussion of computer-based systems.

Human Abstracting Systems

Although automatic abstracting is a popular current topic, all operational abstracting systems are largely *human abstracting systems.* That is, the reading of the original document, selection of portions of it for the abstract, the writing of the abstract and, in many instances, its formatting and editing, are all performed by humans. The most important of these processes, from the viewpoint of the abstract user, is the selection of material from the original document for inclusion in the abstract. But this selection may be influenced considerably by the other processes.

The process of translation from one language to another, as carried out by human abstractors, may result in considerable variation in the size and quality of the abstract. Such variations depend largely upon the abstractor's facility with both the source language and the target language. It seems unlikely that the quality of an abstract prepared by an abstractor with poor facility in one of the languages will be as good as that prepared by one fluent in both. The fault is most serious when the abstractor does not have a good comprehension of the source language. The resulting abstract may be well written grammatically, but may not represent the content of the original document. On the other hand, when the target language is the source of difficulty, there is greater chance that the abstract will contain errors or misinterpretations.

A second factor which influences the selection of material for the abstract is the abstractor's knowledge of the subject area of the document being abstracted. The abstractor who is expert in the subject area will likely produce an abstract which is shorter and more general, and one which requires more knowledge by the reader than an abstract produced by an abstractor whose knowledge of the subject area is marginal. On the other hand, the abstractor with marginal expertise in the subject is perhaps less likely to capture the main thrust of the document in his abstract. The best abstractor is probably one whose qualifications lie between the extremes of expertise and passing knowledge. The following lends credence to this observation: ". . . the best way to get all of the new and significant information of a paper into an abstract, expressed in proper technical terms, is to obtain the spare-time abstracting service of someone actively interested and working in the specific field of chemistry into which a paper being abstracted fits."[12] The suggestion is that it is better (and easier) to teach subject specialists to abstract than to attempt to teach abstractors a subject specialty.

Another factor which influences selection of material for the abstract is the imposition of formatting and editorial responsibilities upon the abstractor. The more mechanical tasks the abstractor has to perform, the greater is the likelihood that the quality of the abstract will suffer. Undoubtedly, Zipf's law of least effort comes into play here.[13] We may suppose that the abstractor of a chemical article would be inclined to use structural drawings rather than words to express, say, a chemical reaction. But if line formulas rather than two-dimensional drawings were demanded, the abstractor would likely prefer words to express the concept. For example, compare three alternate ways of expressing a chemical reaction:[14]

(1) Chlorination of *p*-nitroaniline with potassium chlorate and hydrochloric acid at 25° gave 2,6-dichloro-4-nitroaniline in 87% yield.

(2) $p\text{-}NO_2C_6H_4NH_2 + KClO_3 + HCl \xrightarrow[87\%]{25^\circ} 2,6\text{-}Cl_2(p\text{-}NO_2)C_6H_2NH_2$

(3) 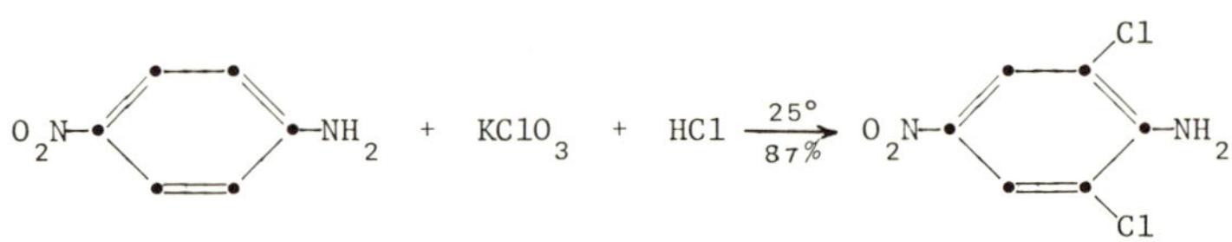

To a chemist, the last example is the easiest to prepare and the easiest to read. Hence, the abstractor should be encouraged to use the language most natural to the field of specialization for which the abstract is to be prepared.

Selection and Consistency

If we ignore these various influences on the nature of the material selected for inclusion in an abstract, we can concentrate our attention on the actual matter of *selection.* How does a human abstractor decide what material in the original document should be included in the abstract? A corollary question is, does the abstractor display any consistency in the selection process (a) within a document, (b) between documents, or (c) through time for either (a) or (b)? In order to answer these questions, one needs to know:

(1) What constitutes a good abstract of a document (or what constitutes a good set of abstracts, if one allows for the possibility of more than one equally good, goal-directed abstract)?

(2) What is the operational definition of an abstract used by a particular abstracting system, as embodied in the rules for abstracting prescribed (or described) by the system,* and how does it contrast with the definition of a "good" abstract?

(3) Given answers to (1) and/or (2), what consistency does an abstractor display in following the rules of the abstracting system or in producing a good abstract (if the rules of the abstracting system appear to be at odds with the abstractor's understanding of what constitutes a good abstract)?

Only the last of these questions has been studied systematically, but no clear answer to it or to any of the other questions has been found. Some of the research done, however, does suggest the direction further efforts should take.

*The directions for preparing author abstracts, issued by a particular journal, constitute an operational definition (no matter how vague) of an abstract.

The most significant studies on selection and consistency are those by Rath, Resnick, and Savage[15] and by the group headed by Edmundson and Wyllys.[16] The main conclusions to be drawn from the Rath-Resnick-Savage studies are that human abstractors show poor consistency in selecting sentences for the abstract both between abstractors and with respect to time. In the Edmundson-Wyllys study of abstractor consistency, it was concluded that, although human abstractors were not very consistent among themselves, the abstracts they produced were adequate to justify a study of the attributes of the sentences they extracted from the document.

Little is known about the process or reasoning through which human abstractors choose particular material from the original article. Neither is it clear to what extent human abstractors are consistent in their abstract production. Perhaps a more important consideration—especially since there is no concrete answer to the question of what constitutes a good abstract—is that questions relating to human selection and consistency may be irrelevant. It is just possible that the abstracts produced by humans are not good. If this is so, then it would be undesirable to try to emulate the human processes, since such emulation would simply lead to a faster rate of production of consistently poor abstracts. Such abstracts might be easy to produce but difficult to sell. The major, unanswered question is: "What makes an abstract a good one?"

Computer-Based Abstracting Systems

Abstracting systems in which the computer plays a central role in the selection process will now be discussed. Again, whether the abstracts produced by such systems are "good" is not known, but, happily, the questions of selection and consistency of selection are easily dealt with. These two aspects of computer-based abstracting systems will be considered in greatest detail in this section.

Eight significant studies in computer-based abstracting have been described in the literature:*

*Other studies might have been included in this list, notably those of Baxendale[17] and of Climenson and Hardwick,[18] but the eight listed are believed to be representative of the various approaches to computer production of abstracts.

(1) The Luhn study.

(2) The ACSI-Matic study conducted by IBM.

(3) The Oswald study.

(4) Word-association research.

(5) The TRW study conducted by Edmundson and Wyllys.

(6) The Earl study.

(7) The Rush, Salvador, and Zamora study.

(8) The Russian work, led by Skoroxod'ko.

Before summarizing the approaches taken in these studies and the nature of the results, it will be useful to describe the basic components of a computer-based abstracting system.

Basic System Components

A computer-based abstracting system must:

read the document to be abstracted;

analyze the document;

apply a set of selection and/or transformation rules to produce the abstract;

format the resulting abstract; and

print the abstract.

Reading the original document is perhaps as difficult a task as any subsequent processing steps the abstracting system must perform. A technical paper will commonly use many different characters in several styles and in a half-dozen fonts. The ordinary computer-system input devices cannot handle such a wide range of characters, nor are there optical readers (optical scanning devices) available which can do so economically, except, perhaps, on a very large scale.[19] One or both of two simple solutions to this problem are usually effected: (1) pre-edit the text, eliminating or altering those portions of the documents which the input devices cannot handle; (2) use an elaborate scheme of flagging of characters which cannot be read directly. Figures, tables, and other graphic materials will in all probability not find their way into the computer at all. Pre-editing is most likely to be used, since the special text features that might otherwise be preserved are of questionable value in the actual abstracting process and since, until the use of optical printers becomes widespread, computer printers cannot economically, if at all, print the necessary range of characters.

Once the text has been read into computer memory, the input text is analyzed. The actual analytical methods used depend upon the particular abstracting system, so a discussion of this component will be deferred until the following sections.

Similarly, rules for selecting those parts of the original document which will be included in the abstract are dependent upon the particular system; these rules will also be considered in the ensuing sections.

Formatting the output of the abstracting process has not been particularly inspired. The usual methods include listing individual sentences or, occasionally, printing the abstract in paragraph format as in the original document. Abstracts produced by humans have never been noted for their literary merit or for their typographic art, so one should not criticize the output of computer-based abstracting systems too severely. Even so, it is not unreasonable to hope that a little imagination might be applied to the formatting of computer-produced abstracts.

No known computer-based abstracting system has utilized any output device other than the line-printer, although certain abstract journals are printed in a two-step process, in the first of which an optical printer (COM, computer-onto-microfilm processor) is used. These publications do not, however, result from the compilation of computer-produced abstracts.

With this general view of the basic processes involved in computer-based abstracting given, specific studies which have lead to the implementation, at least experimentally, of computer-based abstracting systems will be discussed.

The Luhn Study

Luhn is credited with having first suggested and demonstrated that abstracts could be produced through computers.[20] His procedures for generating abstracts by computer were as follows:

(1) The document to be abstracted was first punched into cards (texts which were used required no pre-editing) and then transferred to magnetic tape.

(2) The text was read, word by word. *Common words** were deleted through table lookup. The remaining words, called *content words,* were associated with any punctuation preceding or following them, and their exact location in the original document was noted.

(3) The content words were then sorted into alphabetical order.

(4) Words of similar spelling were "consolidated" as follows: Successive pairs of word-tokens were compared letter-by-letter and at the first point of difference a count was initiated of the number of nonmatches observed from that point to the end of the longer word-token. If this count was less than seven, the word-tokens were taken to be of the same word-type (i.e., to represent the same notion); otherwise the word-tokens were taken to be distinct word-types. The frequency of occurrence of each word-type was then determined and word-types occurring less frequently than some prescribed value were deleted. The remaining word-types were considered to be "significant."

(5) The significant word-types were then sorted into location order.

(6) Next, sentence representativeness was determined. Sentences were divided into substrings, each of which was bounded by

**Common words* might well be called *nonsubstantive* words since these are words that are considered to have no value in determining the significance of a portion (sentence, paragraph, etc.) of text. *Common* should not be confused with *function* in reference to word classification.

significant words separated by no more than four nonsignificant words. Significant words separated from other significant words by more than four words were called "isolated" words and were not given further consideration. For each substring, a representativeness value, r_i, was calculated according to the equation:

$$r_i = \frac{p_i^2}{q_i}$$

The highest r_i for a sentence was taken as its representativeness of document content. Sentences having a value of r_i above a prescribed value (or a predetermined number of sentences of highest r_i) were selected for inclusion in the abstract.

(7) The abstract was then printed as a set of sentences, formatted in paragraph style.

While Luhn's methods for determining word and sentence significance have fallen into some disrepute, the technique is clearly of historical importance. Examples of abstracts produced by these methods, as well as data on word representativeness and word-token and word-type counts, may be found in Luhn's papers.[21]

The ACSI-Matic Study

The ACSI-Matic study, conducted by IBM for the Army Department's *A*ssistant *C*hief of *S*taff for *I*ntelligence, has been the only study to produce a computer-based abstracting system as part of an operational information system.[22]

ACSI-Matic employed selection procedures analogous to those suggested by Luhn. The importance of this study lies in its rather novel variations on Luhn's basic techniques. These modifications included:

(1) Elaboration of Luhn's sentence scoring technique.

(2) Special treatment of documents with an unusually large fraction of low-frequency words.

(3) Special treatment of documents with extraordinarily long sentences.

(4) Choice of sentences, once scored, to form a tentative abstract.

(5) Procedures to reduce redundancy among the sentences selected.

The ACSI-Matic scoring technique improved upon Luhn's treatment of the density of representative words in a sentence. To illustrate the technique, consider the sentence

N R N N R N N N R N N

where N = a nonrepresentative and R = a representative word. R words were given a value of 1 and nonterminal sequences of N-words were given a value of $\frac{1}{2}^n$. Thus, the sentence above would be scored

$$1 + \frac{1}{4} + 1 + \frac{1}{8} + 1 = 3\ \frac{3}{8}.$$

This procedure was applied to sentences of documents whose average sentence length was in the range of 18 to 26 words. When a document had an average sentence length greater than 26 words, each sentence score, computed as above, was divided by the square root of the number of words in the sentence to give a "corrected" score. This procedure slightly favored selection of longer sentences. On the other hand, when more than 10 percent of the sentences of a document were longer than 40 words, the unmodified scoring procedure was employed once again. Thus, the overall effect of these procedures was to give a slight preference to sentences whose lengths were in the range of 26 to 40 words.

Once all sentences in a document had been scored, a set of sentences which were potentially members of the abstract were selected, as follows:

> The number of sentences in the document is divided by ten. If the quotient is more than 20, 20 is substracted from the result and the remainder is divided by 32. The number of sentences in

> the abstract is this quotient plus 20. If the document has fewer than 200 sentences, the abstract has 10 percent of the total number of sentences.[23]

The *n* sentences with the highest scores were designated as "abstract sentences" and the n/4 sentences with the next highest scores were called "reserve sentences." Word-tokens were consolidated by a process analogous to Luhn's method, after which the "abstract sentences" were examined for possible redundancy. Two sentences were considered to be redundant if they contained a number of matching words which was greater than one-fourth of the total number of words compared in the two sentences. Highly redundant sentences were deleted and were replaced by sentences from the "reserve sentence" set. This process was continued until more redundant "abstract sentences" were found or until the set of "reserve sentences" was exhausted. The sentences that were in the set of "abstract sentences" at the end of this process constituted the computer-produced abstract.

The Oswald Study

The essential distinction of the Oswald study is that he employed an indexing criterion in the selection of sentences to form an abstract.[24] This indexing criterion was that groups of words, as well as single words, should be index entries. Consequently, Oswald chose sentences for an abstract that scored high in the number of "representative" word groupings they contained. Representativeness was determined in a manner similar to that employed by Luhn.

Word Association Research

The significance of word-association studies is that they further emphasize the importance of word clusters in conveying an author's intent. Doyle,[25] Bernier,[26] Quillian,[27] and others have attempted to represent a measure of semantic information content (i.e., the measure of the value of some textual structure in conveying an author's intent) through the use of spatial representations of language.

Doyle's association map concept is based on statistical association criteria; a map could, therefore, be produced by computer program.

The method for generating an association map involves the creation of an n x n correlation matrix (n = number of key-words to be correlated) and the correlation of the matrix elements, by means of the Pearson correlation coefficient,[28] for co-occurrence of keywords in individual documents. A portion of a typical association map is shown in Figure 14. Links between nodes (words) show the word associations. The numbers beside the links indicate the strength of the associations, i.e., the values of the Pearson correlation coefficient. Arrows indicate that word co-occurrence is due mainly to a two-word term; the arrows point to the second word of the term.

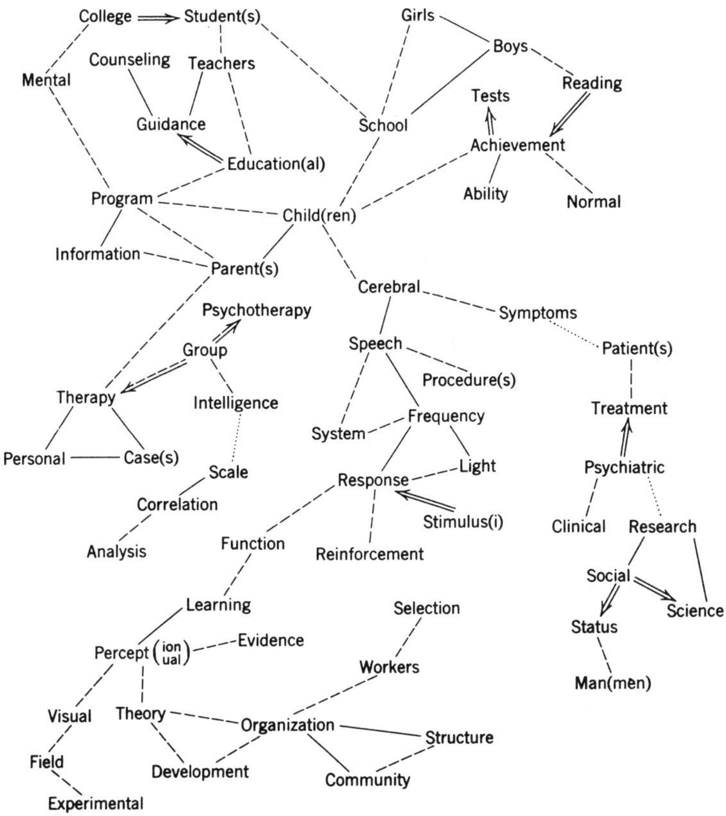

Figure 14–Example of Doyle's Association Map

Doyle envisioned the use of an association map as an associatively organized index that would facilitate the operation of a combined man-machine searching system. But the association map derived from a single document represents a telegraphic abstract in which an attempt has been made to indicate term relations through statistical association.

Quillian's work is superficially similar to that of Doyle, but, in the generation of what one might call "semantic maps," Quillian makes the associations intellectually rather than statistically. These semantic maps make up a memory model which he has employed in studies of human-like language behavior. The significance of this work for abstracting lies in the strong emphasis on *relations* between words (concepts) rather than on the words alone. The details of this work cannot be treated here, but the reader is referred to any of several good papers for this purpose.[29]

Bernier and, more recently, Avramescu[30] and Fugmann[31] have also emphasized the importance of specifying both words and relations between words in indexes and abstracts. Nevertheless, none of them has made a concerted effort to use semantic structure as a basis for creating abstracts of original documents. But since abstracting is fundamentally a matter of semantic content abstraction, it is reasonable to expect that studies along the lines of the investigations mentioned here would be of value in the automated production of abstracts.

The TRW Study (Edmundson and Wyllys)

A major objective of the TRW study in computer-based abstracting was to formulate a method of assigning numerical weights to sentences, so that some of these sentences could be reliably chosen for inclusion in an indicative abstract.

The TRW study developed a logical, mathematical method for the assignment of numerical weights to sentences. It demonstrated the considerable potential of four methods of sentence selection, which were called the Cue, Key, Title, and Location methods (see Figure 15). These methods can be described, briefly, as follows.[32]

The *Cue* method makes use of a list of words, which are classified as bonus words (those that have a positive value or weight in sentence selection); stigma words (words that have a negative weight); and null words (words irrelevant to sentence selection).

		Structural Sources of Clues:	
		Body of Document (Text)	Skeleton of Document (Title, Headings, Format)
Linguistic Sources of Clues:	General Characteristics of Corpus	CUE METHOD: Cue Dictionary (995 words) (Includes Bonus, Stigma, and Null subdictionaries)	LOCATION METHOD: Heading Dictionary (90 words) (Location method also uses ordinal weights)
	Specific Characteristics of Document	KEY METHOD: Key Glossary	TITLE METHOD: Title Glossary

Figure 15–Summary of the Four Methods of Sentence Selection Studied by the TRW Group

The *Key* method is based on the frequency of occurrence of words, and its approach is similar to that used by Luhn in his pioneering study.[33]

The *Title* method is based on a glossary of the words of the title and subtitles (excluding null words). Sentences containing words that also occur in the title are assigned a higher weight than is given either to sentences containing words that also occur in a subtitle, or to sentences with no such words.

The *Location* method is based on the hypothesis that certain headings precede important passages and that topic sentences occur early or late in a document or paragraph. The latter was also Baxendale's hypothesis, resulting from observations made in her studies of automatic indexing.[34]

In the abstracting system developed in this study the relative weights among the four basic methods were parameterized in terms of the linear function

$$a_1 C + a_2 K + a_3 T + a_4 L$$

where a_1, a_2, a_3, and a_4 are the parameters (positive integers) for the Cue, Key, Title, and Location weights, respectively. The Cue-Title-Location method was found to have the highest score, while the Key method in isolation was the poorest of the methods. On the basis of these data, it was decided to omit the Key method as a component in the preferred abstracting process. An example of an abstract produced by use of the combination of the Cue, Title, and Location methods is shown in Figure 16.

The results of the research at TRW indicate that abstracts can be defined, identified, and produced in a computer-based system. The study concluded that future abstracting methods must take into account syntactic and semantic characteristics of the language and the text, and that they could not rely simply on gross statistical evidence. Edmundson concluded that the major task of any further research would be to identify and define the differences between manual and computer-produced abstracts and to minimize these differences so that computer-produced abstracts can supplement, and perhaps compete with, traditional ones.[35]

The Earl Study

The investigation of computer-based informative abstracting and extracting performed by Earl and her associates at Lockheed Missiles and Space Company was aimed at basic research in English morphology, phonetics, and syntax.[36] This study dealt with basic linguistic research as a necessary prerequisite to the computer production of abstracts.

The Rush, Salvador, and Zamora Study

The approach taken by Rush, Salvador, and Zamora to the creation of abstracts by computer is interesting in that it consists mainly of criteria for rejecting sentences.[37] In other studies the emphasis has been on criteria for selecting abstract-worthy sentences. In this study, all sentences of the original document that were not rejected were included in the abstract for that document. Some transformations were applied to the sentence of the extract. This system is also of interest then because the abstracts it produces more closely resemble manually prepared abstracts.

The automatic abstracting system developed by Rush, Salvador, and Zamora consists basically of a dictionary, called the Word Control List (WCL), and a set of rules for implementing certain functions specified for each WCL entry. This combination of rules and dictionary has been designed and implemented to produce indicative abstracts.

The exclusion of sentences from the abstract involves the deletion of words or strings of words which identify sentences giving historical data, results of previous work, examples, explanations, speculative material, and so on. Rush, Salvador, and Zamora claim that a few hundred word strings could serve to eliminate a high percentage of the sentences of a document. Such word strings were incorporated into the WCL.

The WCL consists of an alphabetically ordered set of words and phrases, which are referred to collectively as word strings. The entries in the WCL are treated as functions. Each has two arguments: a *semantic weight* and a *syntactic value.* Each function returns a value which indicates whether the sentence is a candidate for retention or deletion.

PAR SENT DOCUMENT NUMBER 6 PAGE 1

ABSTRACT BASED ON CUE TITLE LOC. WTS.

EVALUATION OF THE EFFECT OF DIMETHYLAMINE BORINE AND SEVERAL OTHER ADDITIVES ON COMBUSTION STABILITY CHARACTERISTICS OF VARIOUS HYDROCARBON TYPE FUELS IN PHILLIPS MICROBURNER (AD87730)

R. L. BRACE

PAR	SENT	
1	0	SUMMARY
2	1	AT THE REQUEST OF THE NAVY BUREAU OF AERONAUTICS, PHILLIPS PETROLEUM COMPANY UNDERTOOK THE EVALUATION OF DIMETHYLAMINE BORINE AS AN ADDITIVE FOR IMPROVING THE COMBUSTION CHARACTERISTICS OF AVIATION GAS TURBINE TYPE FUELS.
2	2	BECAUSE OF THE SMALL AMOUNT (100 GRAMS) OF DIMETHYLAMINE BORINE RECEIVED FROM CALLERY CHEMICAL COMPANY, THIS EVALUATION HAS BEEN LIMITED TO THE MEASUREMENT OF ITS EFFECT ON THE FLASH-BACK CHARACTERISTICS OF THREE PURE HYDROCARBCNS (TOLUENE, NORMAL HEPTANE AND BENZENE) IN THE PHILLIPS MICROBURNER.
3	2	PREVIOUS STUDIES IN PHILLIPS 2 INCH TURBOJET ENGINE TYPE COMBUSTOR HAD INDICATED THAT SUCH MATERIALS COULD SUBSTANTIALLY INCREASE THE MAXIMUM RATE OF HEAT RELEASE ATTAINABLE, ESPECIALLY WITH LOW PERFORMANCE FUELS SUCH AS THE ISO PARAFFIN TYPE HYDROCARBONS-PARTICULARLY WHEN OPERATING UNDER SEVERE CONDITIONS FOR COMBUSTION (I.E. , HIGH AIR FLOW VELOCITY OR LOW COMBUSTION PRESSURE).
4	1	THE ASSUMPTION HAS BEEN MADE IN THIS FUEL EVALUATION THAT THE GREATER THE ALLOWABLE HEAT INPUT RATE FOR A GIVEN VELOCITY, THE GREATER THE DEGREE OF COMBUSTION STABILITY.
4	2	CN THIS BASIS, THE DATA INDICATE THAT ALL THE ADDITIVE MATERIALS TESTED CAUSED AN INCREASE IN STABILITY PERFORMANCE., A FUEL OF RELATIVELY LOW PERFORMANCE SUCH AS TOLUENE BEING BENEFITED TO A GREATER EXTENT THAN A HIGH PERFORMANCE FUEL SUCH AS NORMAL HEPTANE.
4	5	IN GENERAL, ADDITIVE CONCENTRATIONS OF ONE PER CENT BY WEIGHT IN THE SEVERAL PURE HYDROCARBONS WHICH NORMALLY DIFFERED QUITE WIDELY IN PERFORMANCE, PRODUCED UNIFORMLY SUPERIOR COMBUSTION STABILITY CHARACTERISTICS AS MEASURED USING THE PHILLIPS MICROBURNER.
3	0	I. INTRODUCTION
6	1	AT THE REQUEST OF THE NAVY BUREAU OF AERONAUTICS THE JET FUELS GROUP HAS EVALUATED THE EFFECTS OF THE ADDITION OF SMALL AMOUNTS OF DIMETHYLAMINEBORIN =S ON THE COMBUSTION STABILITY PERFORMANCE OF SEVERAL HYDROCARBON FUELS.
7	1	DUE TO THE SMALL QUANTITY OF THIS MATERIAL OBTAINED THE EVALUATION WAS CONDUCTED IN THE PHILLIPS MICROBURNER (MODEL 1A) WHICH IS A SLIGHTLY MODIFIED VERSION OF THE ORIGINAL PHILLIPS MICROBURNER (MODEL 1).

8	0	II. DESCRIPTION OF PHILLIPS MICROBURNER (MODEL 1A)
10	0	III. DESCRIPTION OF TEST APPARATUS
14	0	IV. DESCRIPTION OF TEST FUELS
17	0	V. TEST PROCEDURE
21	0	VI. RESULTS
25	0	VII. DISCUSSION
29	1	PREVIOUS WORK CONDUCTED IN THE PHILLIPS 2 INCH COMBUSTOR (REF. 2) INDICATED THAT SOME ADDITIVES CAUSED A SIGNIFICANT INCREASE IN THE PERFORMANCE OF A LOW RATING FUEL WHILE THESE SAME ADDITIVES DID NOT SUBSTANTIALLY EFFECT THE HIGHER RATING FUELS.
31	3	ALL FOUR ADDITIVES INDICATED THEIR ADDITION TO BE SUBJECT TO THE EFFECT OF DEMINISHING RESULTS UPON FURTHER ADDITION-THAT IS, THEIR EFFECT WAS NOT ESSENTIALLY A BLENDING EFFECT.
33	0	VIII. CONCLUSIONS
34	1	1. THE ADDITION OF DIMETHYLAMINE BORINE IN CONCENTRATIONS OF ONE PER CENT BY WEIGHT TO JET FUEL TYPE HYDROCARBONS RESULTED IN A UNIFORMLY HIGH LEVEL OF COMBUSTION STABILITY PERFORMANCE AS MEASURED BY PHILLIPS MICROBURNER.
35	1	2. THE ADDITION OF RELATIVELY LARGE AMOUNTS OF PROPYLENE OXIDE TO TOLUENE WERE NECESSARY TO PROVIDE SIGNIFICANT IMPROVEMENT IN STABILITY PERFORMANCE AS INDICATED BY INCREASES IN ALLOWABLE HEAT INPUT RATES.
36	1	3. THE ADDITION OF ADDITIVE CONCENTRATIONS (UP TO 1 PER CENT) OF AMYL NITRATE, CUMENE HYDROPEROXIDE, AND DIMETHYLAMINE BORINE ALL RESULTED IN IMPROVED STABILITY PERFORMANCE., THE GREATEST INCREASES WERE SHOWN WHEN BLENDED WITH A FUEL OF POOR PERFORMANCE CHARACTERISTICS-SUCH AS TOLUENE.
37	0	IX. RECOMMENDATIONS
38	1	BASED ON THE EVALUATION OF THE EFFECTS OF ADDITIVES ON THE FLASHBACK LIMITS OF THE ADDITIVE-FUEL BLENDS TESTED IN THE MICROBURNER (MODEL 1A) IT IS RECOMMENDED THAT DIMETHYLAMINE BORINE SHOULD BE FURTHER INVESTIGATED.
38	2	THIS FUTURE WORK SHOULD INCLUDE STUDY OF COMBUSTION STABILITY AND COMBUSTION EFFICIENCY EFFECTS IN THE PHILLIPS 2 INCH COMBUSTOR AND AN INVESTIGATION OF ITS INFLUENCE ON COMBUSTION CLEANLINESS.

Figure 16–An Abstract Produced by the TRW Group Using the Cue, Title, and Location Methods of Sentence Selection

Such a dictionary organization makes it possible to change entries without altering anything else in the system.

The semantic weight of a WCL entry may be either positive or negative; it may cause the sentence to be retained or deleted. Most WCL entries have negative semantic weights, but the system allows a few notable exceptions. Some word strings have semantic clues which indicate that the sentence should be retained. For example, when the author of a document includes in a sentence a word string such as "this paper," "this study," or "present work," it is assumed that these phrases indicate that he is about to say what the document is all about.

The semantic weights from the WCL are implemented using a set of rules. The rules are applied in a fixed order which provides a hierarchy among the various semantic weights. Once a word string has been identified and the semantic weight has indicated whether the sentence should be kept or deleted, its further processing is unnecessary. Flexibility is obtained by incorporating the rules in the program and supplying the semantic weights externally. Thus, the rules could be altered without changing the WCL, and the WCL could be altered independently of the rules.

Since the implementation of the semantic weights requires some syntactic information, a partial syntactic analysis of each sentence is performed. This analysis is carried out through use of the syntactic values associated with entries in the WCL, in conjunction with procedures implemented within the program. The syntactic values are used primarily for making an analysis of the commas in a sentence.

The abstracts obtained from this system appear to be of fair quality. An example of such an abstract is given in Figure 17; other examples can be found in the Rush-Salvador-Zamora article referred to earlier.

The Soviet Study

E. F. Skoroxod'ko has carried out research in automatic abstracting at the Academy of Sciences of the Ukrainian SSR in Kiev, USSR.[38] This research is based on the assumption that only a method of automatic abstracting which adapts itself to a text can provide good and stable results. This assumption is based on the belief that a given

ADDITION OF OXYGEN TO OLEFINS AT LOW TEMPERATURE IV REARRANGEMENTS. #R. KLEIN AND M.D. SCHEER, JOURNAL OF PHYSICAL CHEMISTRY 74(3), 613-616(1970). # A CONSIDERATION OF THE OXYGEN ATOM ADDITION TO CIS- AND TRANS-2-BUTENE IN THE TEMPERATURE REGION 77 TO 113 K LED TO THE FORMULATION OF A NEW TRANSITION INTERMEDIATE. IN THIS INTERMEDIATE, THE OXYGEN ATOM IS REPRESENTED AS BOUND IN A LOOSE, THREE-MEMBERED RING WITH, AND IN THE PLANE OF, THE OLEFINIC STRUCTURE OF THE REACTANT. OBSERVATIONS ON 2-BUTENES HAVE BEEN EXTENDED TO SEVERAL MORE STRAIGHT-CHAIN, INTERNAL OLEFINS IN THE LOW-TEMPERATURE REGION. COMPARISON OF THE TRANS-EPOXIDE TO KETONE RATIONS FROM THE CIS- VS THE TRANS-OLE FIN WITH INCREASING SIZE OF THE OLEFIN INDICATES THAT THESE RATIOS DIVERGE. THE SECOND IS THAT REACTION OF OXYGEN ATOMS IN THE LOW-TEMPERATURE REGION TENDS TO BE MORE STEROSCOPIC WITH TRANS- THAN WITH CIS-OLEFINS. CARBONYL COMPOUNDS CONSTITUTE A SIZEABLE FRACTION OF THE PRODUCTS OF THE OXYGEN ATOM ADDITION TO OLEFINS IN THE LOW-TEMPERATURE REGION AND, AS HAS BEEN NOTED, AN INTRAMOLECULAR GROUP MIGRATION IS REQUIRES FOR CARBONYL FORMATION. THE PRINCIPAL CARBONYL PRODUCT IN THE TRANS-2-BUTENE REACTION AT 90 K IS 2-BUTANONE. THE FORMATION OF THIS KETONE REQUIRES THE MIGRATION OF H. COMPARED TO THE MIGRATION OF THE METHYL GROUP, THAT OF H IS SLIGHTLY FAVORED. CIS-2-BUTENE IS NOT USEFUL FOR THE COMPARISON, AS BOTH OF THE HYDROGEN ATOMS ATTTACHED TO THE OLEFINIC CARBON PAIR ARE SUPPRESSED THROUGH INTERACTION WITH OXYGEN IN THE COMPLEX. THE RELATIVE QUANTITIES OF 2-BUTANONE TO ISOBUTYRALDEHYDE IS TAKEN AS A MEASURE OF THE RATIO OF MIGRATION OF THE HYDROGEN ATOM TO THE METHYL GROUP. REACTIONS WERE EFFECTED AT 90 K IN THE APPARTATUS ROUTINELY USED FOR THIS PURPOSE. THE OLEFINS WERE DILUTED 10 TO 1 WITH PROPANE. THE EXPOSURE TIME OF OXYGEN ATOMS WAS 5 MINUTES, AND ABOUT 1% OF THE OLEFIN WAS REACTED. THE PRODUCTS WERE DETERMINED AT 135 AND A HELIUM FLOW OF 100 CC/MINUTE. THE CIS AND TRANS ISOMERS OF 3,4-EPOXY-3,4-DIMETHYLHEXANE WERE NOT SEPARABLE. LOCALIZTION OF THE OXYGEN ATOM IN THE TRANSITION COMPLEX PRECEDING ALKYL GROUP REARRANGEMENT IS NOT IN ACCORD WITH THE EXPERIMENTAL RESULTS. AT 90 K, THE RATIO OF ADDITION TO C-2 IS 16 TIMES THAT TO C-3. FOR MEP, ADDITION OF THE OXYGEN ATOM TO THAT CARBON ATOM OF THE DOUBLE BOND TO WHICH THE TWO METHYL GROUPS ARE ATTACHED WOULD BE EXPECTED TO BE FAVORED.

Figure 17–Abstract Produced by the System Developed by Rush, Salvador, and Zamora

text has individual characteristics which should determine the optimal selection criteria and abstracting procedure.

The individual characteristics of a given text are defined by the form of a semantic network which represents the text. A semantic network is defined to be a graph where the nodes are associated with sentences and the arcs are associated with semantic relations between sentences. Two sentences, A and B, are said to be semantically related if (1) at least one

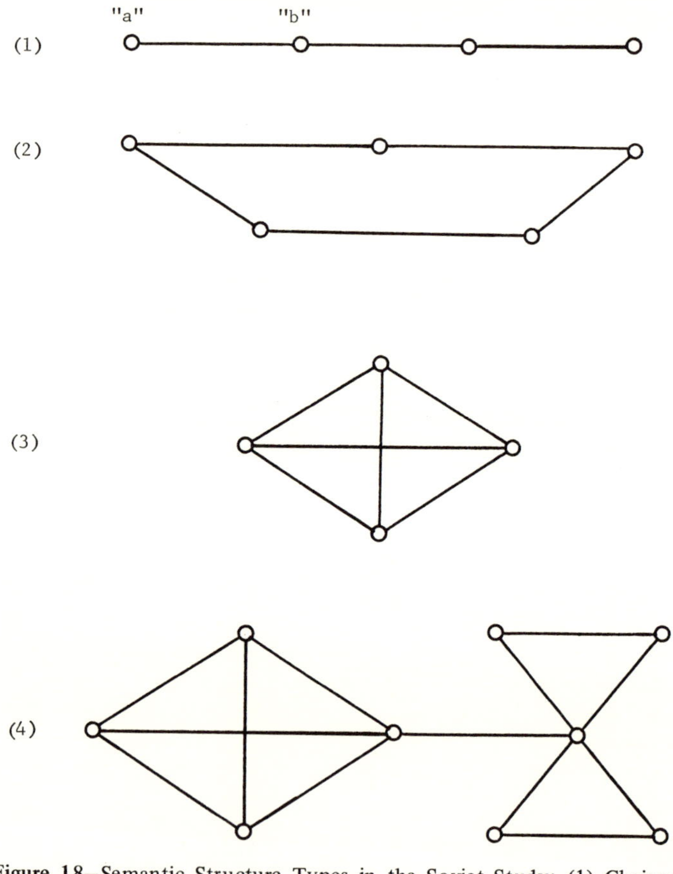

Figure 18–Semantic Structure Types in the Soviet Study: (1) Chained, (2) Ringed, (3) Monolithic, and (4) Piecewise

noun occurs in both sentences A and B, or (2) sentence A contains a word "a" and sentence B contains a word "b," where "a" and "b" have been predefined as being semantically related, or (3) words "a" and "b" in sentences A and B are related with respect to a given text. The graphs of the semantic structures can be identified in terms of four important types (see Figure 18). The selection of an appropriate abstracting procedure is based on the type of semantic structure.

The significance of each sentence is assumed to be directly proportional to the number of sentences semantically related to it. Thus, nodes in the graph that have the most incident arcs are defined to be the most significant. The sentence significance also depends upon the amount of change in the semantic network for that document when the node for that sentence is removed from the network.

Based on this measure, Skoroxod'ko concludes that in the case of chained or ringed structure networks, it is impossible to form an adequate extract from sentences taken from the text since all sentences have approximately the same semantic value. Automatic abstracts can only be generated from texts where semantic relationships can be depicted as monolithic or piecewise structures.

CONCLUSION

In concluding this discussion of computer-based abstracting, it is noteworthy that no system so far developed is yet ready for use in a production environment. The reasons for this, we believe, are two-fold: excessive cost and rather poor quality of the product. In all fairness, however, it should be pointed out that the main concern of the designers of the various systems discussed here (save, perhaps, for the ACSI-Matic system) was to test various ideas for computer-based abstracting. Hence, efficiency and cost are not easily treated in a meaningful way.

The quality of abstracts produced can be treated, if not more meaningfully, at least more fairly. The problems which have plagued researchers in computer-based abstracting, as well as in all other areas of text processing, may all be thought of as language problems. That is, the problems are either of a semantic (interpretive) or syntactic (physical) nature. To deal with these problems effectively, a much better knowledge of language must be gained. In the meantime we can expect progress in automatic abstracting to be slow and the development of production systems to be several years in coming.

NOTES

1. "Abstracting," *Encyclopaedia Brittanica,* Vol. 1 (Chicago: 1957), pp. 67-68.

2. B. Weil, "Standards for Writing Abstracts," *Journal of the American Society for Information Science* 21, no. 5 (1970): 351-357.

3. American Chemical Society, *Handbook for Authors of Papers in the Journals of the American Chemical Society,* American Chemical Society Publication, (Washington, D.C., 1967).

4. Chemical Abstracts Service, *Directions for Abstractors,* The Ohio State University, Columbus, Ohio, 1971 Revision.

5. R. E. Wyllys, "Extracting and Abstracting by Computer," in H. Borko (ed.), *Automated Language Processing* (New York: John Wiley and Sons, 1968).

6. R. E. Maizell, J. F. Smith, and T. E. R. Singer, *Abstracting Scientific and Technical Literature* (New York: Wiley-Interscience, 1971).

7. H. Borko and S. Chatman, "Criteria for Acceptable Abstracts: A Survey of Abstractors' Instructions," *American Documentation* 14, no. 2 (1963): 149-160.

8. J. E. Rush, R. Salvador, and A. Zamora, "Automatic Abstracting and Indexing. II. Production of Indicative Abstracts by Application of Contextual Inference and Syntactic Coherence Criteria," *Journal of the American Society for Information Science* 22, no. 4 (1971): 260-274.

9. B. A. Mathis and J. E. Rush, "Abstracting," *Encyclopedia of Computer Science and Technology,* J. Belzer, A. G. Holzman and A. Kent (eds.), (New York: Marcel Dekker, 1973).

10. *A System Study of Abstracting and Indexing in the United States,* System Development Corporation, Falls Church, Virginia, 1966 (PB 174 249).

11. Mathis and Rush, "Abstracting."

12. E. J. Crane (ed.), *The Production of Chemical Abstracts* (Washington, D.C.: American Chemical Society, 1959).

13. G. K. Zipf, *Human Behavior and the Principle of Least Effort* (Cambridge, Mass.: Addison-Wesley Publishing Company, 1949).

14. From P. P. T. Sak and S. A. Peoples, *Arzneimittel-Forschung* 11 (1961): 27-33.

15. G. J. Rath, A. Resnick, and T. F. Savage, "The Formation of Abstracts by the Selection of Sentences. Part I. Sentence Selection by Men and Machines," *American Documentation* 12, no. 2 (1961): 139-141; and A. Resnick, "The

Formation of Abstracts by the Selection of Sentences. Part II. The Reliability of People in Selecting Sentences," *American Documentation* 12, no. 2 (1963): 141-143.

16. H. P. Edmundson, V. A. Oswald, and R. E. Wyllys, *Automatic Indexing and Abstracting of the Contents of Documents,* Planning Research Corporation, Los Angeles, California, prepared for Rome Air Development Center, Air Research and Development Command, Griffiss Air Force Base, New York, 1959 (AD 231 606); and *Final Report on the Study for Automatic Abstracting,* Thompson Ramo Wooldridge, Inc., Canoga Park, California, 1961 (PB 166 532).

17. P. B. Baxendale, "Machine-Made Index for Technical Literature–An Experiment," *I.B.M. Journal of Research and Development* 2, no. 4 (1958): 354-361.

18. W. D. Climenson, N. H. Hardwick, and S. N. Jacobson, "Automatic Syntax Analysis in Machine Indexing and Abstracting," *American Documentation* 12, no. 2 (1961): 178-183.

19. *Special Report on Optical Character Recognition,* Auerbach Corporation, Philadelphia, Pennsylvania, 1971; and R. G. Casey and G. Nagy, "Advances in Pattern Recognition," *Scientific American* 224, no. 4 (1971): 56-71.

20. H. P. Luhn, "The Automatic Creation of Literature Abstracts," *I.B.M. Journal of Research and Development* 2, no. 2 (1958): 159-165.

21. Ibid., and H. P. Luhn, *An Experiment in Auto-Abstracting,* Auto-Abstract of Area 5 Conference Papers, International Conference on Scientific Information, Washington, D.C., I.B.M. Research Center, Yorktown Heights, New York, 1958.

22. I.B.M. Corporation, Advanced Systems Development Division, *ACSI-matic Auto-Abstracting Project, Final Report,* Vols. 1 and 3, Yorktown Heights, New York, 1960 and 1961.

23. *Special Report on Optical Character Recognition,* Auerbach Corporation.

24. Edmundson, Oswald, and Wyllys, *Automatic Indexing.*

25. L. B. Doyle, "Semantic Road Maps for Literature Searchers," *Journal of the Association for Computing Machinery* 8, no. 4 (1961): 553-578; and "Indexing and Abstracting by Association," *American Documentation* 13, no. 4 (1962): 378-390.

26. C. L. Bernier and K. F. Heumann, "Correlative Indexes III. Semantic Relations Among Semantemes–The Technical Thesaurus," *American Documentation* 8, no. 3 (1957): 211-220.

27. M. R. Quillian, "The Teachable Language Comprehender: A Simulation Program and Theory of Language," *Communications of the Association for Computing Machinery* 12, no. 8 (1969): 459-476.

28. Doyle, "Indexing and Abstracting by Association."

29. M. R. Quillian, *A Notation for Representing Conceptual Information: An Application to Semantics and Mechanical English Paraphrasing,* SP-1395, System Development Corporation, Santa Monica, California, 1963; "Word Concepts: A Theory and Simulation of Some Basic Semantic Capabilities," *Behavioral Science*

12, no. 5 (1967): 410-430; and "Semantic Memory," in M. Minsky (ed.), *Semantic Information Processing* (Cambridge, Mass.: The M.I.T. Press, 1968).

30. A. Avramescu, "Probabilistic Criteria for the Objective Design of Descriptor Languages," *Journal of the American Society for Information Science* 22, no. 2 (1971): 85-95.

31. R. Fugmann, H. Nickelsen, I. Nickelsen, and J. H. Winter, "TOSAR–A Topological Model for the Representation of Synthetic and Analytical Relations of Concepts," *Angewandte Chemie International Edition in English* 9, no. 8 (1970): 589-595.

32. H. P. Edmundson, "New Methods in Automatic Extracting," *Journal of the Association for Computing Machinery* 16, no. 2 (1969): 264-285.

33. Luhn, "The Automatic Creation of Literature Abstracts."

34. Baxendale, "Machine-Made Index for Technical Literature."

35. Edmundson, "New Methods in Automatic Extracting."

36. L. L. Earl, "Experiments in Automatic Extracting and Indexing," *Information Storage and Retrieval* 6, no. 4 (1970): 313-334.

37. Rush, Salvador, and Zamora, "Automatic Abstracting and Indexing"; and R. Salvador, *Automatic Abstracting and Indexing,* The Computer and Information Science Research Center, The Ohio State University, Columbus, Ohio, 1969, Technical Report Series (OSU-CISRC-TR-69-15).

38. E. F. Skoroxod'ko, "Information Retrieval System of the Ukrainian Institute of Cybernetics," *International Congress on Scientific Information,* Moscow, 1968; and "Adaptive Method of Automatic Abstracting and Indexing," *IFIP Congress 71,* Ljubljana, Yugoslavia, 1971, Booklet TA-6, pp. 133-137.

5

General Principles of Retrieval Systems

Specialists in information theory have shown that a general communication system takes the form shown in Figure 19. All communication takes place through these basic steps. A source produces a message, a transmitter sends the encoded message via some sort of signal along a channel, and a device receives the signal and then converts it to the original message for delivery at the destination. "Noise" is any disrupting influence acting on the signal as it traverses the channel; it may distort or obliterate the message if there is too much of it.

Perhaps the most interesting thing about Figure 19 is the fact that it does seem to be general. That is, it applies to a communication system consisting of two people shouting down a hall just as much as it does to a telephone system or to two-way radio communication. A system for information storage and retrieval can be thought of as a special case of the general communication system where the processes are especially independent of time. The ancient Greeks wrote much that is as relevant and interesting to us today as the daily news on television. The fact that this material is transmitted to us via the printed word centuries after its first appearance does not matter; we are involved with a communication system just as real as the one which broadcasts our news to us instantaneously. It is food for thought to realize that there is a direct

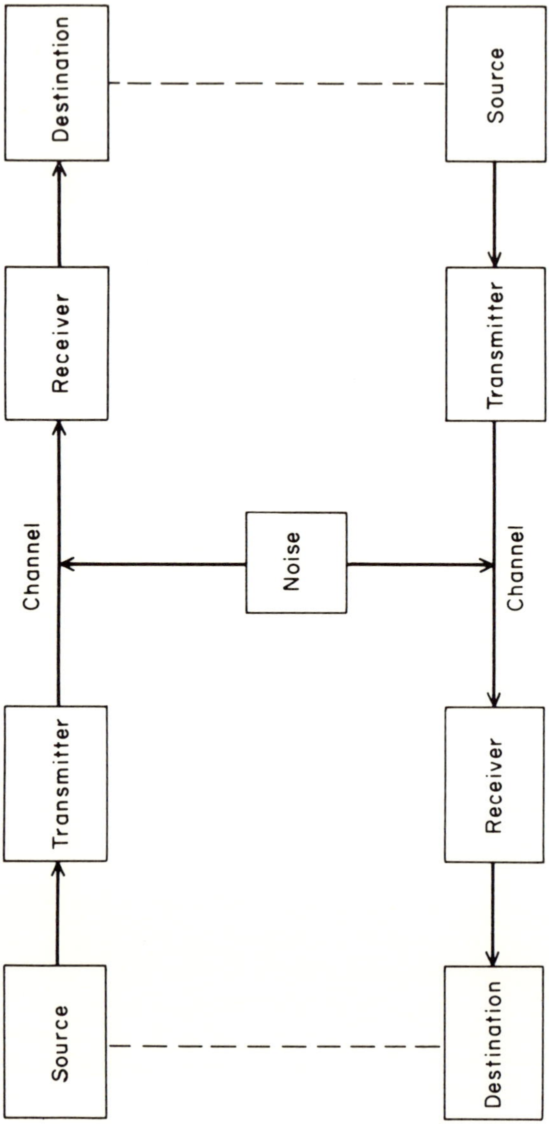

Figure 19—Model of a General Communication System

analogy between static and "snow" on radio and television and the errors which creep into the editorial and printing processes; it does not require much imagination to think of many other examples of "noise."

INFORMATION STORAGE AND RETRIEVAL

As a proper subset of communication, information storage and retrieval can be subjected to further analysis. When people speak of "information retrieval," they usually mean document retrieval; this is natural, because information is stored in documents that are then indexed and/or classified to facilitate their retrieval. Figure 20 shows a flow diagram of an information storage and retrieval system for documents. New documents being input to the system are indexed according to the vocabulary control device employed (if any) and are added to the data base. The vocabulary control device will typically be one of the syndetic structures mentioned in Chapter 3: a list of subject headings, a thesaurus, or one of the specialized classification schemes suitable for subject analysis. During the indexing process new terms may be discovered which should be added to the vocabulary control device. Indexing should not be thought of as merely a passive activity that consists of looking up preferred terms from a source presumed infallible. Ideally, it should be a dynamic process involving continual improvement of whatever syndetic structure is used.

Information or document retrieval is the reverse side of this coin. Search strategy is formulated in terms of the vocabulary used by the system being interrogated, and output is obtained from the system in the form of a subset of the data base. Since inadequacies in the vocabulary control device may also be detected at this stage, the formulation of search strategy should be considered a dynamic process too. The results of a good search should not be too large to allow rational examination of the output, nor should they be so small as to exclude documents containing relevant information. Achieving the desired results by formulating optimal search strategy is often as much art as science, but there are definitely some guidelines of universal validity which will help the searcher. Guidelines and other helpful hints for the

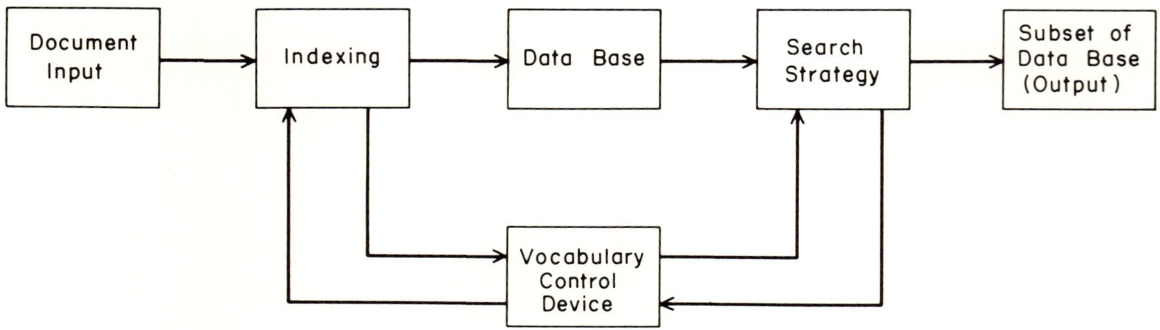

Figure 20–A Flow Diagram Depicting an Information Storage and Retrieval System for Documents

searcher using typical mechanized systems are covered later in this chapter.

TYPES OF SEARCH SERVICES

Regardless of whether they are performed manually or with the help of mechanical devices, libraries and information centers provide basically two types of search services: retrospective and current-awareness services.

Retrospective searches are simply literature searches made on the entire collection at hand, and they are meant to be comprehensive within the parameters of the search. Although they may sometimes be restricted to a certain time period, more often they seek to be exhaustive by finding all relevant documents from the earliest to the most recent.

Actually, two kinds of current-awareness services are in common use, but only one of them may properly be called a search service. Many libraries and information centers distribute copies of their accessions lists, which are lists of all their recent acquisitions. This is a relatively cheap way of accomplishing the objective, but if the lists are very long this method creates a real inconvenience to the user, who must wade his way through the entire listing to make sure he has found everything he wants. A much more satisfactory method is SDI (*S*elective *D*issemination of *I*nformation), which alerts the user to recent acquisitions only in those areas in which he has expressed an interest. This technique is much more sophisticated and cannot be provided manually for any large number of users where the rate of acquisitions is high. Such an SDI service can be implemented using the techniques of the retrospective search. That is, the search strategy employed for a given user's retrospective search can be used to search through the recent acquisitions every so often. However, care must be taken to provide for updating this strategy, because the user's interests can be expected to change from time to time.

SEARCH LOGIC

It is useful to think of the collections of information centers and libraries or, for that matter, of any given data bases as sets of items.

This allows one to apply elements of set theory and Boolean algebra in analyzing the information system and in formulating the best search strategy for interrogating it. The basic theory involved is the same whether the system is computer-based or uses such manual devices as edge-notched cards; it deals with the coordination of data elements within the records comprising a file.

Strictly speaking, a set is any collection of items that satisfies the following three conditions: (1) each item must have an attribute or attributes which allow one to say "yes" or "no" to the question of whether the item is a member of the set; (2) each item must be distinct and should not appear more than once; and (3) the order of the items in the set should be inconsequential (this is similar to the distinction between combinations and permutations). Most data bases satisfy these conditions and are therefore subject to set operations and the basic laws of Boolean logic.

A file of documents is composed of records that represent the documents, and the records in turn are made up of data elements that usually serve to describe either the physical nature of the documents or their subject content. The data elements may or may not be arranged by fields within each record, and these fields may be of either fixed or varying length. Most people associate the coordination of data elements only with the coordination of the unit terms or subject index terms which may have been used to describe the contents of the documents being sought. However, any data element within the record may be subject to the coordination techniques involved in the search logic, including such nonsubject elements as author name, length of the document, or any other physical attribute of the document, provided they have been included in the record.

Boolean Search Logic

There are several ways of discussing and representing set operations and basic Boolean logic. The best-known of these, including graphic representations and the corresponding notation schemes, are summarized in Figure 21. This figure is a variation on a table prepared by Brandhorst.[1]

	AND	OR (Inclusive)	OR (Exclusive)	NOT
COMMON TERMS	Conjunction	Disjunction	Disjunction	Complement
	Intersection Multiplication Product	Union Sum Combination	Union Sum Combination	Negation Difference
COMMON NOTATION	$A \bullet B$	$A + B$	$A \oplus B$	$A - B$
	$A \cap B$	$A \cup B$	$A \not\cup B$	$A\overline{B}$
	$A \wedge B$	$A \vee B$	$A \not\vee B$	AB'
	$A * B$			
	$A \times B$			

VENN DIAGRAMS

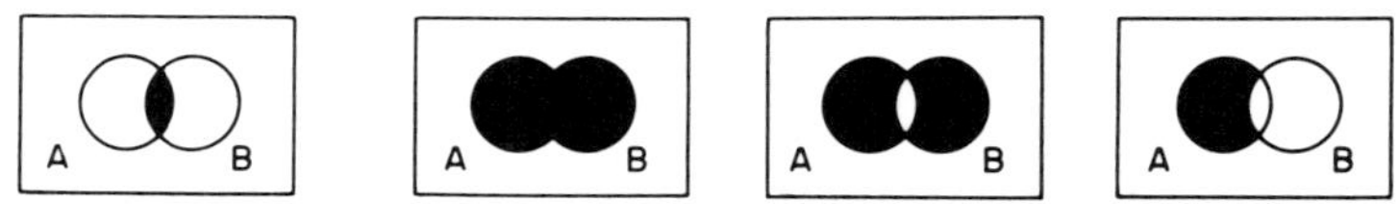

TRUTH TABLE
(1 = True, 0 = False)

A B	AND	OR (Inclusive)	OR (Exclusive)	NOT
0 0	0	0	0	0
0 1	0	1	1	0
1 0	0	1	1	1
1 1	1	1	0	0

Figure 21–Common Forms of Representation of Boolean Logical Operations

Naturally, many operations are not illustrated in Figure 21, but those which are shown are definitely familiar in information retrieval. Also, many more than two data elements may be involved in any given search. For example, a search involving conjunction (Logical AND) among three terms or data elements could be represented by the following Venn diagram:

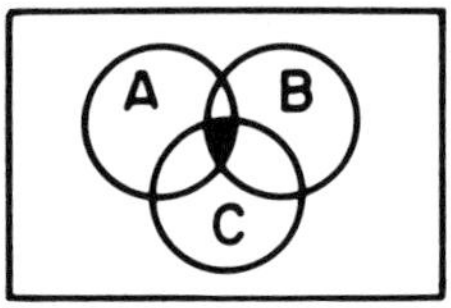

The rectangle stands for the universe of documents in the data base, and the circles represent those sets of documents in the collection which have been indexed under the terms (or other data elements) A, B and C. The shaded part of the diagram corresponds to those documents which have been indexed under all three terms.

One further theoretical aspect is that the circles in Venn diagrams need not intersect. When they do not intersect, it means that the subset of documents described by the intersection is null; i.e., if one asks for all those documents indexed under terms A and B and these two sets are disjoint, then no documents will satisfy the search logic and none will be retrieved.

Although abstractions such as "Term A" are being used to keep the discussion general, it should be clear that one can write search logic dealing with any one of a variety of situations relating to the real world. For example, perfectly legitimate search logic could be written for 'carbon disulfide' *or* 'nitrobenzene' *not* 'solvent' if one were interested in those documents dealing with the two chemicals as reagents. In most systems, inserting "*not* 'solvent' " would not guarantee the relevance of all documents retrieved, but it would help minimize the chances of getting unwanted material.

Retrieval systems vary considerably in complexity. Some will allow only simple coordination among a very few terms; others will permit arbitrarily complex logical operations. It is not unusual to find coordination of the type shown in Figure 22.

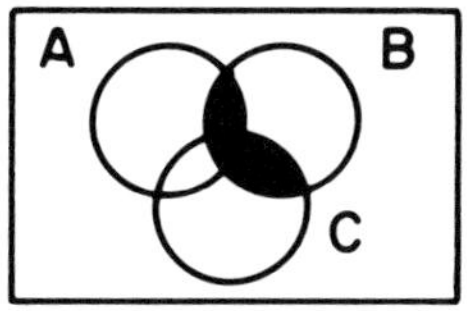

Figure 22–Venn Diagram for the Relationship (A · B) + (B · C)

Weighted-Term Search Logic

Weighted-term search logic is similar to Boolean search logic and shares the same theoretical basis in set theory. However, it is more flexible in that it permits the searcher to assign each term a number which is proportional to the value that the searcher places on that term for the particular search being undertaken.

To use a typical weighted-term system, it is first necessary to select a threshold value or cutoff weight. This value is an arbitrary integer within the limits of the system, and it remains constant throughout each individual search. Weights relative to this threshold value are assigned to each term used in a given search. If the weight assigned to a term is equal to or greater than the threshold value, then the computer will cause the information associated with documents indexed under that term to be presented to the searcher. If the weight assigned to the term is less than the threshold value, then the information associated with the documents indexed under the term will not be produced unless the term appears in combination with one or more other terms and the sum of the terms' weights is equal to or greater than the threshold value. This allows for very flexible coordination of the terms in the search strategy. Consider the following example:

Threshold value = 5

Term	Weight
A	4
B	1
C	1
D	1
E	1

In this case the documents indexed under term A will be retrieved only if the documents have also been indexed under terms B, C, D, or E, or any combination of them.

Negative weights can also be assigned to terms. This feature is analogous to logical negation and is valuable in minimizing the chances of retrieving unwanted material. The next example illustrates negative weighting:

Threshold value = 7

Term	Weight
A	5
B	8
C	−6

Documents indexed under term B will be printed out only if term C has not been used also. If the document has been indexed under both terms A and B, then it will be retrieved even if it has also been indexed under term C, because $13 - 6 = 7$, which is the threshold value.

As a final example, weighted-term logic for the relationship shown in Figure 22 is given below:

Threshold value = 5

Term	Weight	or	Term	Weight
A	1		A	2
B	4		B	3
C	1		C	2

Remember that the choice of the threshold value is arbitrary. A number other than 5 could be used with equally satisfactory results, provided that the proper weights were assigned to the terms.

Interestingly, negative threshold values have their uses too. If one wants all documents indexed under term A or term B but not both, then a negative threshold value can be declared, and the weights of terms A and B can be set equal to it. If either of the desired terms appears by itself, then the document will be retrieved since the value of the term will equal that of the threshold value. However, if both terms appear, then the document will not be produced because their combined weight will be less than the threshold value. This is the weighted-term equivalent of the Boolean exclusive OR as shown in Figure 21.

The sum of the weights of the terms under which a document has been indexed can be thought of as a kind of document weight. This number is usually provided along with the information about the document, and serves as an approximate measure of the document's probable relevance to each search request. Some weighted-term systems provide a sort on these document weights, so that the documents will be provided in decreasing order of their probable relevance. In addition, such systems often permit the searcher to specify an upper limit to the number of documents which a search may yield. Thus, if several hundred documents are likely to satisfy a given search request, then an arbitrary cutoff number of, say, 50, could be assigned to limit the output to 50 documents. If there has been a sort by document weights, then the searcher will be looking through 50 of the most promising documents rather than a random sample.

Another feature of some of the more sophisticated weighted-term systems is the built-in capability of determining from the document weight precisely which terms were responsible for its being produced. This is not always easy or even possible using small integers because different combinations of these numbers will not produce unique results. Some of these systems employ powers of two as term weights; others allow a combination of Boolean and weighted-term capabilities. Interested readers are referred to the appropriate articles in the notes section at the end of this chapter.[2]

Term Truncation

Because many terms often have the same roots, it is sometimes useful to match on portions of the terms. For example, a searcher might be interested in obtaining information about the acetyl group regardless of whether the group occurred, for example, in an acid, an aldehyde, an amide, or in compounds like acetoacetic acid. In such a case, it would be useful to be able to search for a pattern match on the character string 'acet.' This is made possible in many systems by the introduction of a special character, often an asterisk or a slash mark, which indicates where the pattern match should stop. Thus, a searcher could write the term 'acet*' or 'acet/' in his search logic and coordinate it with any other data elements. The logic of the search would be as in all preceding examples, but any term beginning with 'acet' would produce a successful pattern match. Naturally, this would serve to broaden the search and would result in many more hits.

Although it is more common to truncate on the end of a term, it is sometimes useful to be able to truncate at the beginning, as in the example of a searcher wanting everything on '____mycin.' Systems permitting this kind of truncation will usually employ special symbols at the front of the term as before, and they may also allow truncation on both ends simultaneously. Such uses of truncation are particularly important because they provide one means of performing substructure searches, which allow the user to query the system with respect to moieties present within molecular structures. Substructure searching is more common in systems using the linear notations discussed in Chapter 8 and the topological systems covered in Chapter 9 because the notation employed is usually more rigorous than typical chemical nomenclature and also, as in the case of linear notation, because of the advantages of brevity.

Despite its value to the searcher, truncation can prove troublesome if the stem used as a search term is not chosen with care. For example, the term 'carbox____' might correspond, in a typical data base, to such words as carboxyl, carboxylase, carboxylate, carboxylating, carboxylative, carboxylic, and so on, which could lead to the retrieval of unwanted documents. Truncation on both ends of a term may be still more difficult to manage properly. For these reasons, syndetic devices have been developed to aid the searcher in the proper choice of truncated search terms.[3] The most popular of these devices is the KLIC (*K*ey-*L*etter-*I*n-*C*ontext) index developed by Kent (see Figure 23).

GUIDELINES AND HELPFUL HINTS FOR THE SEARCHER

There are some useful rules which will guide the searcher wishing to revise his search strategy. In the simplest case, involving a single term or

ST	ACHYBOTRYS
T	ACHYCARDIAS
TR	ACHYLOBANE
T	ACHYPHYLAXIS
	ACHYRANTHES
TR	ACHYSAURUS
BR	ACHYURAN
AST	ACI
DIPS	ACI
AC	ACIA
ARB	ACIA
PIST	ACIA
ENCEPHALOMAL	ACIA
GL	ACIAL
SP	ACIAL
INTERF	ACIAL
EM	ACIATION
GL	ACIATION

Figure 23–Example of the KLIC Index

descriptor, if undesired output results, then the searcher must choose either another term or a more complex search strategy requiring a coordination technique. If the search logic which produced unsatisfactory output already involved coordination techniques, then the searcher needs to consider the types of coordination as well as the terms or descriptors used. If the search has resulted in too little output, it is possible that there has been excessive use of logical AND or NOT, since both of these operators tend to restrict the search. Similarly, if there has been too much output, then there may have been too much use of logical OR or excessive use of term truncation, both of which tend to broaden a search. Counteracting these tendencies may involve cutting back on the use of the offending type of operator or introducing an operator having the opposite effect. Naturally, the searcher may wish to change the terms used in a search as well, but first it would be wise for him to examine how he coordinated the ones already tried.

At this stage it should probably be pointed out that zero output is not necessarily indicative of a bad search. Consider, for example, the case of an inventor wishing to establish the novelty of a new device and his likely reaction to finding no patents for similar devices, or the case of the doctoral student who would be delighted to find no Ph.D. dissertations already written on his chosen topic.

Finally, one should be wary of a search producing 100 percent relevant output; something may very well have been missed because the search was too narrow.

CHOICE OF DEVICES AND MEDIA

Of the devices which permit coordination and facilitate information retrieval, the computer is certainly the most glamorous and awesome. However, there are simpler and much cheaper devices which work remarkably well for files of modest size. Perhaps the most common of these are the edge-notched and optical coincidence card file systems.

Edge-notched cards come in a variety of sizes, but basically they consist of cards with holes around their periphery. The interior of these cards can be used for typing, writing, or otherwise storing data, perhaps by using microfilm to make what is called an aperture card. Their chief advantage over ordinary index cards is that information can be retrieved regardless of the sequence of the cards in the file. Storage and retrieval are accomplished by notching the cards, that is, by physically removing the material between the holes and edges of the cards. These notches correspond to a predetermined code, and when a needle is run through a deck of these cards, those that have been notched will fall off while the rest will remain on the needle. Some edge-notched cards have more than one row of holes. As a result of such an arrangement, deep notches from the innermost holes to the edge of the card can be made, and slots can be cut which connect one hole with another. When a needle is run through a deck via an inner hole, the cards that have received deep notches will fall off the needle, while those that have been slotted will simply change their position relative to the cards which have neither been notched nor slotted. Thus, the deck of cards can be separated into as many as three classes by a single needle sort.

The process of notching or slotting edge-notched cards can be thought of as a binary process. That is, a card is either notched or it is not; it is either slotted or it is not. In this respect, the cards can be compared to the main memory of a digital computer, since the individual bits in the memory are either "off" or "on." The binary process is really the same in each case and can be used to represent "yes" or "no" as responses to questions formulated using Boolean logic.

The most straightforward method of coding an edge-notched card is direct coding, in which there is a one-to-one correspondence between each hole and the descriptor it represents. However, when there are more descriptors than there are holes, one must use a method of coding which is more economical from the standpoint of space on the card. For example, two or more holes can be used in combination to represent one term, and since a great number of permutations and combinations are possible on even an ordinary edge-notched card, many more terms or descriptors can be coded. Unfortunately, more needling operations become necessary, and one finds that he has traded efficiency for space on the card. The chief disadvantage of edge-notched cards is mechanical: needles can only be so long and still be practical. Therefore, edge-notched card files are used whenever the collection of documents is relatively small. Moreover, since most people have only two hands, the implementation of search logic involving functions of more than two variables becomes quite cumbersome. Nevertheless, proper coordination of the needles allows for the simple Boolean operations of AND, OR, and NOT. A typical edge-notched card is shown in Figure 24.

Optical coincidence cards are known by a variety of other names. Among these are 'aspect,' 'peek-a-boo,' and the trade name 'Termatrex.' Optical coincidence cards differ from edge-notched cards in that they have holes drilled in the interior of the cards while the written or printed matter appears somewhere on the periphery. Usually, when optical coincidence cards are used for document retrieval, each card in the file represents an indexing term or descriptor. This term is written or printed on the periphery of the card, and the documents in the collection are represented by coordinates on a grid in the center of the card. If a document in the collection has been indexed under the particular term on that card, then a hole is drilled at the coordinates

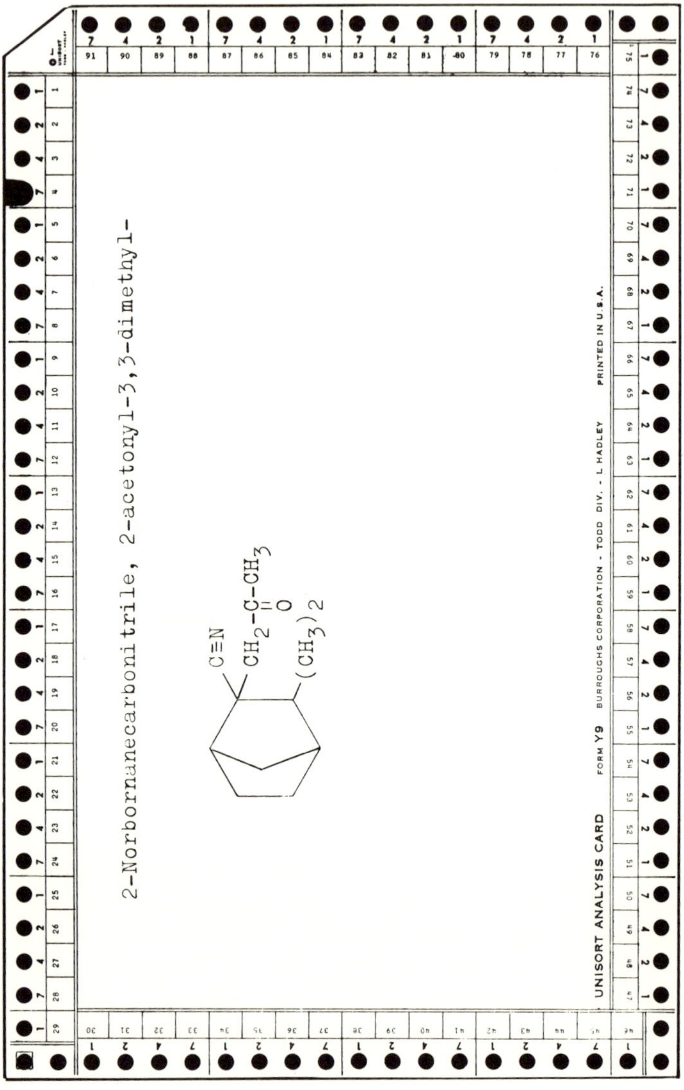

Figure 24—A Typical Edge-notched Card

representing that document. By placing the card over a light source, one can easily identify the documents indexed under that term by looking for the spots where the light shines through. In addition, it is possible to superimpose two or more cards so that the light will shine through only those holes which represent documents indexed under both or all of the terms on the cards superimposed on the light source (logical AND). The drilling of holes in an optical coincidence card is analogous to the notching or slotting of edge-notched cards in that this, too, represents a binary process which can be used to provide "yes" or "no" responses to questions phrased in Boolean logic. All the standard Boolean operations can be achieved using optical coincidence cards, but the methods may be a bit cumbersome, requiring the searcher to keep records of coordinates and to relate them to a separate file of documents or document surrogates. A representative optical coincidence card is shown in Figure 25.

The principal advantage of optical coincidence cards over edge-notched cards is that more documents can be coded conveniently. An optical coincidence card usually has enough positions for 10,000 holes (a 100 x 100 matrix), each of which would represent a document. Since it is the cards which represent documents in an edge-notched card file, a comparable file would contain 10,000 cards. Assuming there are about 100 cards per inch, the edge-notched card file containing 10,000 cards would be around 100 inches long, thus requiring either a nine-foot needle or several passes with a needle of conventional length—clearly a nuisance. On the other hand, optical coincidence cards require special equipment for precision drilling of holes and an appropriate light source. Thus, while some edge-notched card equipment more sophisticated than a needle does exist, it is not a necessity, whereas the equipment associated with the optical coincidence cards definitely is, and it is substantially more expensive.

The main reason for the expense associated with typical optical coincidence systems is the high coding density used. As mentioned above, 10,000 positions are common for optical coincidence cards. An area approximately 6 x 6 inches is used to contain these code positions; this spacing is quite close and requires tolerances to be quite small. If, however, such high densities are not required (i.e., a smaller file is to be

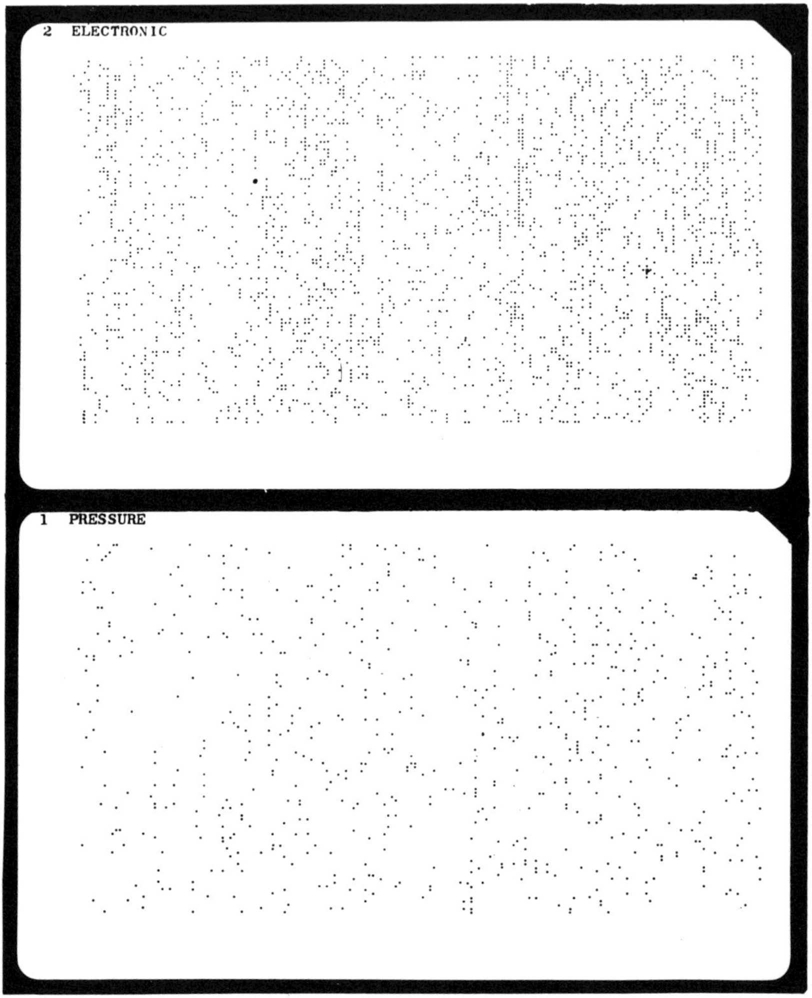

Figure 25–Optical Coincidence Cards

indexed), then a simpler system may be devised using Hollerith cards. The only special equipment needed now is a keypunch (possibly a hand punch). Again, terms are recorded at the top of the card (typed or handwritten) and a punch is made in the card at the coordinates corresponding to the document number. The maximum number of documents which can be represented on a single card is 960 (80 columns x 12 rows); hence, this type of optical coincidence card is useful mainly for small files. A representative set of cards is shown in Figure 26.

Although manual card files may not represent the wave of the future in information retrieval, they can be extremely useful for small personal collections and for specialized files within existing libraries and information centers.[4]

Since computer-based systems are inherently expensive because of the equipment used, it is especially important to consider how best to optimize the system design, both from the standpoint of the organization of the data and the choice of devices and media. With respect to the organization of the data, perhaps the single most important consideration is file organization: the question of how the records comprising the file can best be organized to facilitate retrieval. Although a detailed discussion of file organization is beyond the scope of this book, the reader should at least be acquainted with some of the elementary principles.

There are basically two types of file organization: inverted and uninverted. The optical coincidence card file system discussed earlier is a good example of an inverted file. It is one in which all the document numbers have been "posted" to the terms or descriptors under which they have been indexed. The edge-notched card file, on the other hand, is an example of an uninverted file, where the order of the documents or their surrogates is independent of the terms under which they are indexed. Because of their enormous flexibility, computer-based systems can handle either of these basic organizations and several variations on them.

Ideally, a computer-based system should have all the data stored in the main memory since this would provide the fastest possible access. Unfortunately, the state of the art is such that main memories are far

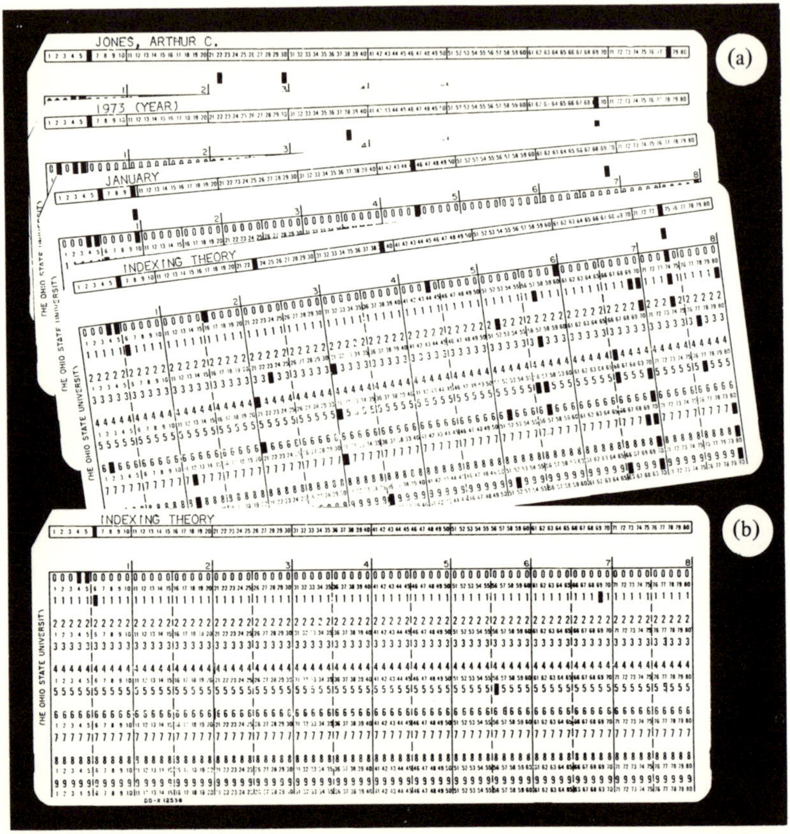

Figure 26—Use of Hollerith Cards as the Basis of an Optical Coincidence System

A document number is represented as a punch in a particular column and row (e.g., 0169). (a) four term cards; (b) the four cards overlapped.

too small for most data bases, requiring that some sort of auxiliary storage device be used. If the data base is set up in a sequential fashion, then magnetic tape, which is a linear, sequential-access medium, may be adequate. Magnetic disk storage may be preferable when one wants direct access to "clusters" of data and is a *sine qua non* for on-line, real-time systems. In any case, the choice of file organization and the format of individual records should be decided upon only after consultation with a systems person experienced in problems of computer-based information retrieval.

Although it is theoretically possible to record the entire contents of all the documents in a collection and to process them by computer, it is seldom done in practice. In a collection of even moderate size, the amount of material to be stored would be staggering, and it would require an unconscionable amount of time to print the results of most searches. For these reasons, output from the computer usually appears in one of three modes: (1) accession or call numbers only, (2) accession or call numbers with complete or partial bibliographic citations, or (3) accession or call numbers with an abstract or annotation in addition to the foregoing. Since accession or call numbers alone do not provide much information, they are usually not distributed by themselves. Rather, a master file of citations and/or abstracts is kept on hand so that copies can be made to supplement the search. This procedure minimizes the amount of computer time required but assumes that good copying equipment and plenty of clerical help are available.

Finally, it must be noted that regardless of the form of the output of a search there should be adequate provision for feedback from the user of the system. Only then can user satisfaction be ascertained. Information should be solicited with respect to the relevance of the data provided and to the form and mode of its provision. Feedback from the user can be invaluable in evaluating the performance of the retrieval system, which is the subject of the next chapter.

NOTES

1. W. T. Brandhorst, "A Table of Set Theory Notations," *Journal of the American Society for Information Science* 21, no. 6 (1970): 428.
2. W. T. Brandhorst,. "Simulation of Boolean Logic Constraints Through The

Use of Term Weights," *American Documentation* 17 (1966): 145-146; Howard P. Iker, "Solution of Boolean Equations Through Use of Term Weights to the Base Two," *American Documentation* 18 (1967): 47; F. W. Matthews and L. Thomson, "Weighted Term Search: A Computer Program for an Inverted Coordinate Index on Magnetic Tape," *Journal of Chemical Documentation* 7 (1967): 49-56; John Mulvihill and Everett H. Brenner, "Ranking Boolean Search Output," *American Documentation* 19 (1968): 204-205; and Helen G. Sommar and Don E. Dennis, "A New Method of Weighted Term Searching with a Highly Structured Thesaurus," *Proceedings of the American Society for Information Science* 6 (1969): 193-198.

3. A. K. Kent, "The Chemical Society Research Unit in Information Storage and Retrieval," *Svensk Kemisk Tidskrift* 80, no. 2 (1968): 39-46; and D. S. Colombo and J. E. Rush, "Use of Word Fragments in Computer-Based Retrieval Systems," *Journal of Chemical Documentation* 9, no. 1 (1969): 47-50.

4. For a good discussion of manual card systems, see Charles P. Bourne, *Methods of Information Handling* (New York: John Wiley & Sons, 1963), pp. 80-115; or R. S. Casey, J. W. Perry, A. Kent, and M. Berry (eds.), *Punched Cards: Their Application to Science and Industry*, 2d ed. (New York: Reinhold, 1958).

6

Retrieval Systems Evaluation

In this book we have explored many ways in which information can be represented, organized, and retrieved. A question of paramount importance that has still to be considered is: "How good or useful are the various information storage and retrieval (IS&R) systems?" This question will be explored in this chapter.

We have all had occasion to use IS&R systems of one sort or another, either of the manual or automated variety, to seek some desired information, and we have all experienced various degrees of elation or dejection or uncertainty as a consequence of our efforts. We experience *elation* if we are fortunate enough to find what we were searching for; *dejection* if we are not; and *uncertainty* if we are unable to tell whether or not we have found what we wanted.

For instance, suppose a person is in search of information on some chemical reaction of current interest to him. He may attempt to find the desired information in any of several alternative, possibly compatible, ways. Printed indexes are a good starting point. A search of printed indexes requires a bit of ingenuity, but one also has the opportunity to use his intuition to its fullest potential. If the desired information is not found in this way, the searcher may turn to an automated system for help. Here, too, ingenuity is called for, but the searcher's intuition is likely to be stifled, partly because of the somewhat less flexible nature of the system and partly because of the "driver-in-the-passenger's-seat" syndrome. These problems are lessened–if not altogether obviated–when interactive automated IS&R systems are used.

If the desired information is not found through the use of available IS&R systems, one may turn to friends and colleagues or else conclude that the information does not exist. However, it is conceivable that the information was not found because of faulty indexing or other factors which are not evident to the searcher. With experience, a searcher soon learns that a "no" answer to a search request frequently causes a lower confidence in the result than he would like. In fact, considerable uncertainty about the availability of the desired information nearly always remains when a "no" response is received.

It should be evident from these arguments that a successful search for an item of information will always require less overall effort than an unsuccessful search. We must qualify this statement, however. If a request is made for *all* information of a given type, the search effort, as well as the confidence in the result, will be the same regardless of the outcome of the search. One or two simple examples may clarify these issues.

Suppose a request is made for the boiling point of *m*-bromoethylbenzene. Recourse might first be had to a familiar handbook, such as the *Handbook of Chemistry and Physics,* Beilstein's *Handbuch der organischen Chemie,* or Huntress and Mulliken's *Identification of Pure Organic Compounds.* If the desired information is found, the search is terminated and is deemed successful. If the boiling point is not found, a more comprehensive index to the chemical literature, such as the Subject Indexes to *Chemical Abstracts, Index Chemicus,* or *Chemisches Zentralblatt,* will likely be used. Again, if the boiling point is found, the search is terminated. Otherwise, the search continues until these information resources are exhausted. When all available information resources have been tried without producing the desired information, the search is terminated and is termed unsuccessful. But it is not certain whether lack of success stemmed from the nonexistence of the information, or whether the information was missed in the search. The overall search process described here is depicted in Figure 27.*

*The situation described is purely hypothetical. One might well terminate the search because of fatigue or disgust.

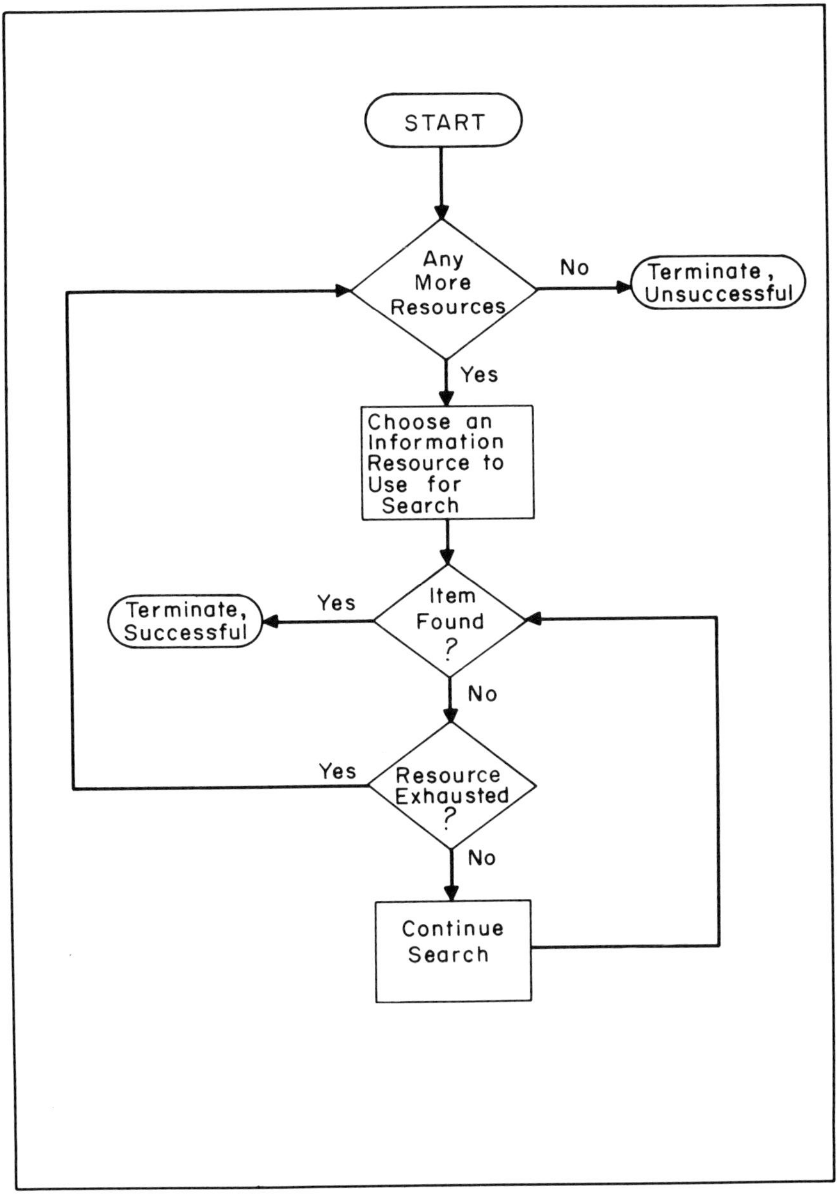

Figure 27–General Model Depicting the Search for a Specific Item

On the other hand, suppose the search request was for *all* information on *m*-bromoethylbenzene. For this type of request the discovery of an item of information does not cause the search to be terminated as it did in the previous example. Only when all available resources have been tried can the search be terminated. But whether the search is counted as successful or unsuccessful, a shadow of doubt always remains as to its efficacy. Searches of this type may be depicted as in Figure 28.

There is no direct evidence to indicate how frequently searches are carried out according to the schemes of Figures 27 and 28, but it is likely that most proceed along the lines of Figure 27. These kinds of searches are performed every day in every chemical laboratory. The type of search illustrated in Figure 28 is typified by a "prior art" search surrounding a patent application. The United States Patent Office, for instance, must carry out exhaustive searches both to determine the validity of the application and to ascertain its scope. Only in the case of the discovery of a "duplicate" patent or of one which "anticipates" the invention at the heart of the application would a nonexhaustive search be performed. A positive proof is always better than a negative one.

THE NOTION OF EVALUATION

Let us now return to the question posed at the beginning of the previous section, namely: "How good or useful are the various IS&R systems?" Many people have attempted to answer this question without, we believe, much success. The troublesome aspect of "evaluation" is that there are too many different criteria upon which to base evaluative procedures and none of these criteria has, or will likely receive, wide acceptance. This is because evaluation always reduces, ultimately, to a personal value judgment.[1]

With traditional information resources, such value judgments have always been made, but the benefit of these judgments has seldom been available to IS&R system operators except through a rise or decrease in sales. While the financial standing of a concern operating an IS&R system is a good reflection of the collective value judgments of the system's users and potential users, from the system operator's view-

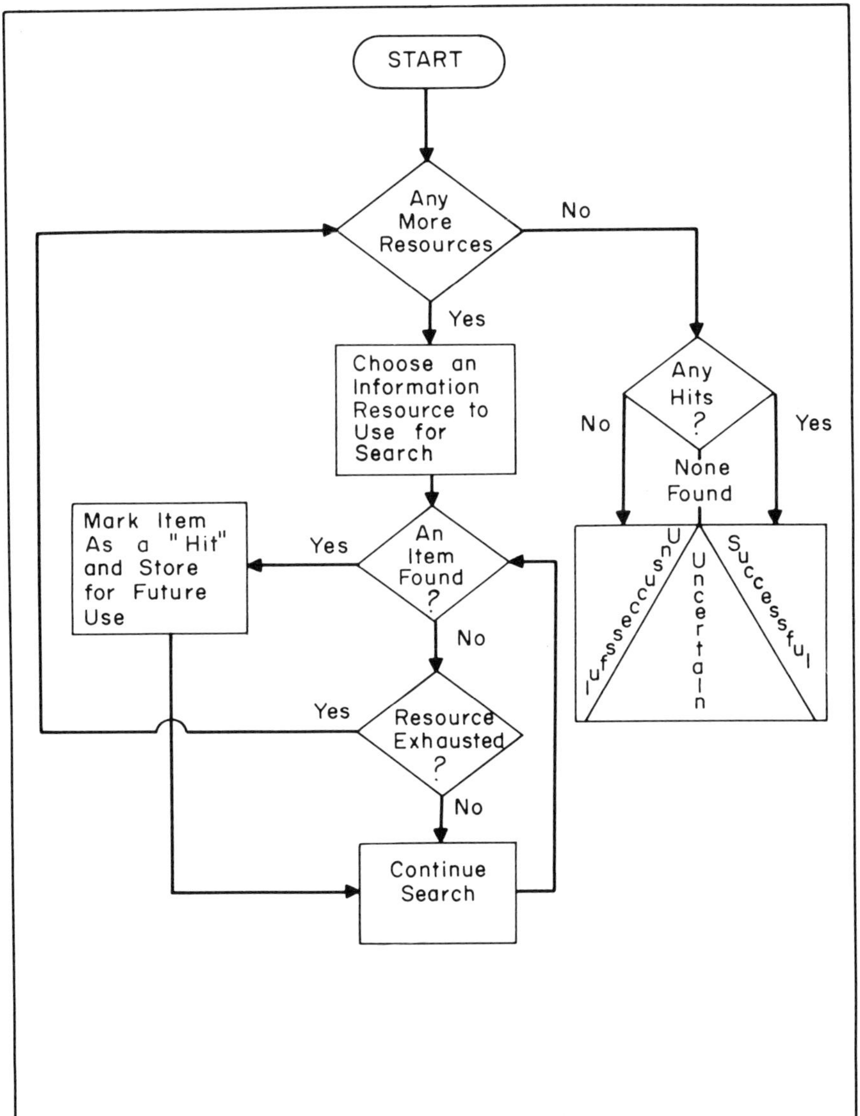

Figure 28–General Model Representing the Search for All Items of a Given Type

point it is an undesirable criterion of evaluation. It is only the worst sort of dullard who has to be forced out of business in order to be shown that his product has need of improvement. It is for this reason and because of a genuine desire to provide good service that IS&R system operators have sought a better method to evaluate their systems.

With the advent of mechanized IS&R systems, a more positive feedback mechanism was introduced. This often took the form of a card that the user filled out, indicating which of the retrieved items were of use to him and which were not. Other comments may also have been provided, whether solicited or not. This approach soon gave rise to a method of evaluation that was of a seemingly mathematically precise nature. If each user indicated which of the documents the IS&R system provided him were of "interest" and which were not, the system operator ought then to be able to say something about the performance of his system. The system operator interpreted "interest" to mean "relevant" or "good" or "useful," and went about computing a percentage of the documents which the system had retrieved as relevant documents. This percentage, frequently called the precision of a search, is expressed

$$\text{Precision} = \frac{\text{Number of relevant documents retrieved}}{\text{Number of documents retrieved}}$$

System operators were not, however, content with this "measure." They were also concerned that something of interest to a user might have been missed in a search. So a measure of the number of documents of interest which were missed in the search was sought. This was a more difficult measure to apprehend, but it was recognized that if a user could himself search the entire collection of documents he could determine if any had been missed. Thus, the notion of *recall* arose. Recall is expressed

$$\text{Recall} = \frac{\text{Number of relevant documents retrieved}}{\text{Number of relevant documents in the system}}$$

These "measures" seem to have had their genesis in the early 1960s.[2] They have continued to have only moderately diminished popularity ever since, despite a few serious attacks on their validity.[3] Let us consider what these "measures" signify.

PRECISION AND RECALL

When first considered, precision seems to be a sound notion and one that would be relatively easy to obtain values for. The total number of documents retrieved by an IS&R system is easily ascertained, and the searcher (user) ought to be able to distinguish between relevant and nonrelevant documents. But is relevance a strictly binary criterion? We think not. It seems unreasonable to suppose that every document retrieved in response to a search request is either perfectly relevant or completely nonrelevant. In fact, it is much more reasonable to suppose that relevance may take on values from a continuum and that precision is not as easily determined as the expression given earlier would suggest. A somewhat more reasonable expression for precision is the following.

$$\text{Precision} = \frac{1}{n} \sum_{i=1}^{n} \text{relevance of document i},$$

where n is the number of documents retrieved and where $0 \leq \text{relevance} \leq \infty$. Regardless of the form of expression, precision is not a particularly satisfactory measure of the performance of an IS&R system. This is partly borne out by the fact that most operating IS&R systems find that 40 to 50 percent precision is usually satisfactory.

If precision is difficult to determine, *recall* is practically impossible; otherwise, there can be no reason for achieving less-than-perfect results. Consider what the recall expression tells us. First, it tells us that we can determine the relevance of each document retrieved and can produce two classes of documents: those that are relevant and those that are nonrelevant. Next, it tells us that we can find all the relevant documents in the system store. This is an absurd conjecture. If we can find all relevant documents in the store, then we must be able to determine relevance unerringly. If these conjectures are so, then we can always obtain perfect recall. One is thus lead to the conclusion that either an IS&R system yields less than perfect recall out of sheer perverseness, or that the notion of recall is useless.

The consequences of these arguments are that recall has no practical significance and that precision is on uncertain grounds. Thus, we need some better ways of evaluating IS&R systems. Some alternative techniques will be suggested later, but first the various factors important in system evaluation will be explored more thoroughly.

FACTORS TO BE CONSIDERED IN SYSTEM EVALUATION

As was pointed out in Chapter 5, an IS&R system incorporates several component processes, each of which contributes to the success or failure of the system. These processes include the following:

Document selection and acquisition. This factor establishes the scope of the system (i.e., the subject area), regardless of claims made for the system.

Document storage. This factor is of little importance to a user when the system does not supply a document in response to a query (few do). Libraries typify the systems for which this is an important factor.

Document representation. Here is perhaps the most important factor governing the effectiveness of any IS&R system. Accurate and complete document representation is essential for the proper functioning of an IS&R system. It is through such representations that a user's query and the document store are brought into coincidence.

Search request formulation and processing. This factor concerns the ability of the IS&R system to accept user queries, formulate a search strategy, and process this search strategy against the document representations already provided.

Formulation and issuance of search responses. An otherwise well-designed IS&R system may be judged to be of low value because of ill-conceived output facilities.

Cost/benefit performance. An otherwise well-designed IS&R system may be judged to be of low value because the cost of using it is greater than the benefit derived.

It should be possible to establish separate evaluation criteria for each of these factors, some being more readily quantifiable than others. It is also desirable to find a way of combining these criteria meaningfully, although knowledge of the performance level of each component separately may prove sufficient.

Document Selection and Acquisition

An IS&R system cannot be any better than the documents it acquires. Thus, considerable attention should be paid to the establishment of working criteria for selecting documents to be incorporated into the IS&R system. Document selection is either based on specific criteria which define the scope (subject coverage) of the system, or else selection defines the scope by default. Document selection also governs the level of detail which can be expected from the system. Finally, the reputation (a value judgment) of the document source must be considered. The reputation of an IS&R system hinges on that of its document collection. In spite of the importance of this factor to the performance of IS&R systems, no studies of note have been undertaken which deal with document selection and acquisition. Paisley and Parker, however, have recognized that source deficiencies may severely limit system performance.[4]

Document Storage

Unless the IS&R system provides a document in response to a query,* evaluation of this factor remains internal to the system. Most IS&R systems provide only an accession number or a bibliographic citation in response to a user's request, which the user employs to recover the document from some other system. A few systems provide both abstracts and bibliographic citations as query responses. In either case, that system's document store is of little concern to the user. The library is the only IS&R system in which the user has an interest in the document store. The business of the library is to satisfy the patron's need for some document, without regard for how he came to know of the existence of that document. That the library might also serve to

*The Institute for Scientific Information's Original Article Tear Sheet (Oats®) Service is one instance of such a system.

help the patron discover the existence of a document is clearly ancillary to the function of document storage. (Several projects are underway which would make the library a truly complete IS&R system.[5]) The document given to the patron need not be the original, but only a facsimile, such as a microform copy. Of course, in the rare instances when the patron asks to see the original, such as an ancient manuscript, original letter, or rare book, his wishes should be met.

Evaluation of the document storage factor can be considered from two points of view (not wholly compatible). The first of these is reflected in the system's ability to produce a document on demand. It is presumed that the document exists and that its existence gave rise to the request. It must also be presumed that the document exists in the system store (otherwise we would be evaluating document acquisition). Systems which are known to possess a document that cannot be produced upon request are very annoying. While the reasons why a document cannot be produced are many and varied (lost, stolen, not reshelved, out to another patron, in the bindery, etc.), they do not lessen the patron's frustration. Thus, a way of evaluating the document storage factor would be to maintain statistics on the number and kind of transaction and to weigh unsatisfied patron requests against the successfully processed requests. Transactions that terminate successfully could be evaluated on the basis of the time required to complete the transaction, while transactions that terminate with failure could be evaluated on the basis of time, reason, and percent of total transactions. To our knowledge there have been no detailed studies of document storage from this point of view, except perhaps for the Long study.[6]

Document storage can also be evaluated from the point of view that it allows for patron *browsing*. That is, can the patron search through the document store without having to pose a formal request (for something he probably can't express anyway) to the IS&R system? Closed-stack libraries, as well as systems that do not provide documents at all, clearly do not satisfy a user's need for browsing. On the other hand, permitting the patron to browse in the document store may cause difficulty in providing documents on demand (through loss, misplacement, etc.). Nevertheless, it is likely that the system which allows browsing of its document store will be judged better than one that does not. With the advent of on-line, computer-based IS&R systems, especially those which employ video-display devices, browsing can be permitted without the associated dangers of disarrangement of the document store.

Document Representation

As already emphasized, the evaluation of IS&R systems must be centered on the needs of its users. These needs are especially critical when considering document representation. The purpose of document representation is to make it possible, or more facile, to bring users and documents into correspondence. One of the main problems is that very often document representations are created without any knowledge of potential system-user needs. Yet, the way in which the user communicates with the system and the way in which the system responds to user requests are dictated almost entirely by the document representation. Consequently, the evaluation of document representations must yield data that can be used to decide which changes in the document representation will provide improved communication with the user.

A wide range of techniques is available for representing the contents of documents (see Chapters 3 and 4); some of these techniques are provided by the document itself through its title, literature citations, and section headings. For the most part, however, the various representations of which a document is capable are provided by the IS&R system. In order to determine the best representation, or set of representations, for a document, the IS&R system must know its users. This is difficult to do. For specific discipline-oriented IS&R systems, however, a first approximation to such knowledge can be gained by employing system designers experienced in the discipline. This insures that typical thought and behavior patterns of potential clients are reasonably accounted for.

Analysis of a sample of the documents to be represented may also give clues as to the best forms of representation. In particular, it will provide an indication of how users of the proposed system may express their wishes and how they would want responses from the system to be formulated. In short, document representations must be provided which will enable users to get what they want in as convenient a way as possible, commensurate with their ability to pay.

Once the IS&R system has established criteria and procedures for document representation, these must be evaluated to make certain that they remain viable. But how are such evaluations to be carried out, and upon what criteria are they to be based?

In order to attempt an answer to these questions, we must examine the central feature of all IS&R activity: *the need of some person for information.* Landry has recently considered the concept of information need in considerable detail.[7] Briefly, *information need* arises from a user's need to acquire data with which to test some hypothesis he has formulated about the nature of his world. Thus, a person comes to an IS&R system with the intent of discovering data that will satisfy his current information need. It follows, therefore, that the purpose of the IS&R sytem is to support this discovery process. But this implies an interaction, even if indirect or slow, between system and user. Carbonell has conceptualized this interaction in somewhat the manner shown in Figure 29.[8] In this figure, the user's memory (data structure) interacts with the IS&R system data structure (document representation) through the user's evaluation, decision-making, and control functions. The environment influences all components of the interface. The user's interaction with the IS&R system will be guided by hypotheses he has formulated about how the data are to be retrieved from the system. The nature of these hypotheses will depend on many things, the more important factor being the user's knowledge of the system's structure. For instance, the library is not a system to which one who knows its structure will likely put a request directly for the solubility of progesterone in water. It is not that such information does not reside in the document store of the library; it is simply that the document representations provided by the library do not permit direct retrieval of such a specific item. If one were unaware of the structure of the library, he might well make such a request; he would then receive data which would cause him to revise his hypotheses.

Supposing, however, that an effective interaction between user and system has been established, the user then directs this interaction toward the satisfaction of his information need. As mentioned earlier,

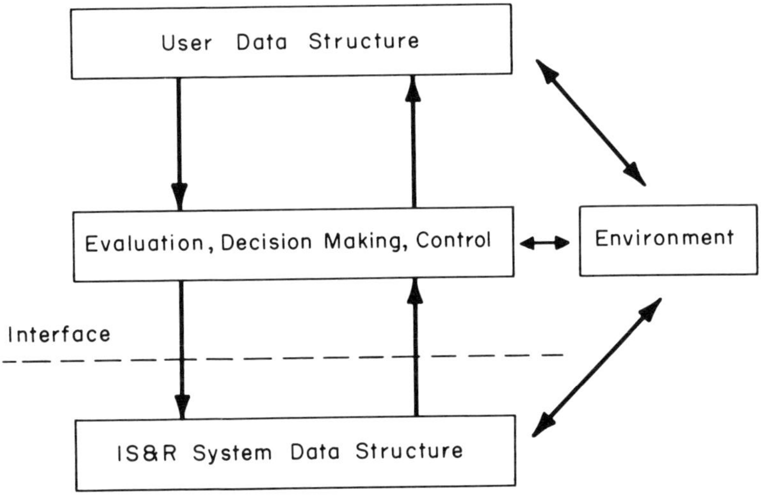

Figure 29–Conceptualization of the Interaction Between a User and an IS&R System (after Carbonell)

this information takes the form of a hypothesis about the user's world which is to be tested by acquiring data from an IS&R system. Acquiring this data is what the interaction between user and system is all about. Evaluation of the interaction is based on the criterion of satisfaction of this information need. The cyclic nature of the process of information retrieval is shown in Figure 30.[9] In this figure, it is assumed that the process begins with information need. The user expresses this need (at least internally) in terms of a structure or pattern of acceptable retrieved data. (It is this factor which enables a user to tell whether some data are acceptable, even when he is unable to say how he knows.) His need for information is made known to the system through a query prepared in a manner suited to the system, with the aid perhaps of system language facilities such as authority lists and thesauri. The system presents the user with some retrieved data (possibly none) or else data *meta* to the request (term not in vocabularly; there are 504 documents in the store which satisfy the request) which the user must then process in some way. If the retrieved data fit into his structure of acceptable data, then he must determine if his information need is satisfied. If it is not—which is the most likely case for the first round or two with the system—the user must modify his structure of acceptable data and try again. If his need is satisfied, the user then reinitiates the information retrieval process as soon as a new information need arises.

Several points can be made about this interaction relative to evaluation of document representation. First, it is clearly document representation that is at the heart of the system side of the interaction. Hence, satisfaction of a user's information need hinges directly upon adequate document representation. In other words, for any given user, that document representation will be judged adequate that satisfies his information need (subject to the provisos to follow).

Second, the effort required of the user in query preparation is at least indirectly dependent upon document representation. A user's judgment of the value of a system is likely diminished in direct proportion to the amount of work he has to do to make his request acceptable to the system. It is for this reason that many systems currently in operation provide either manual or automated assistance in query formulation.

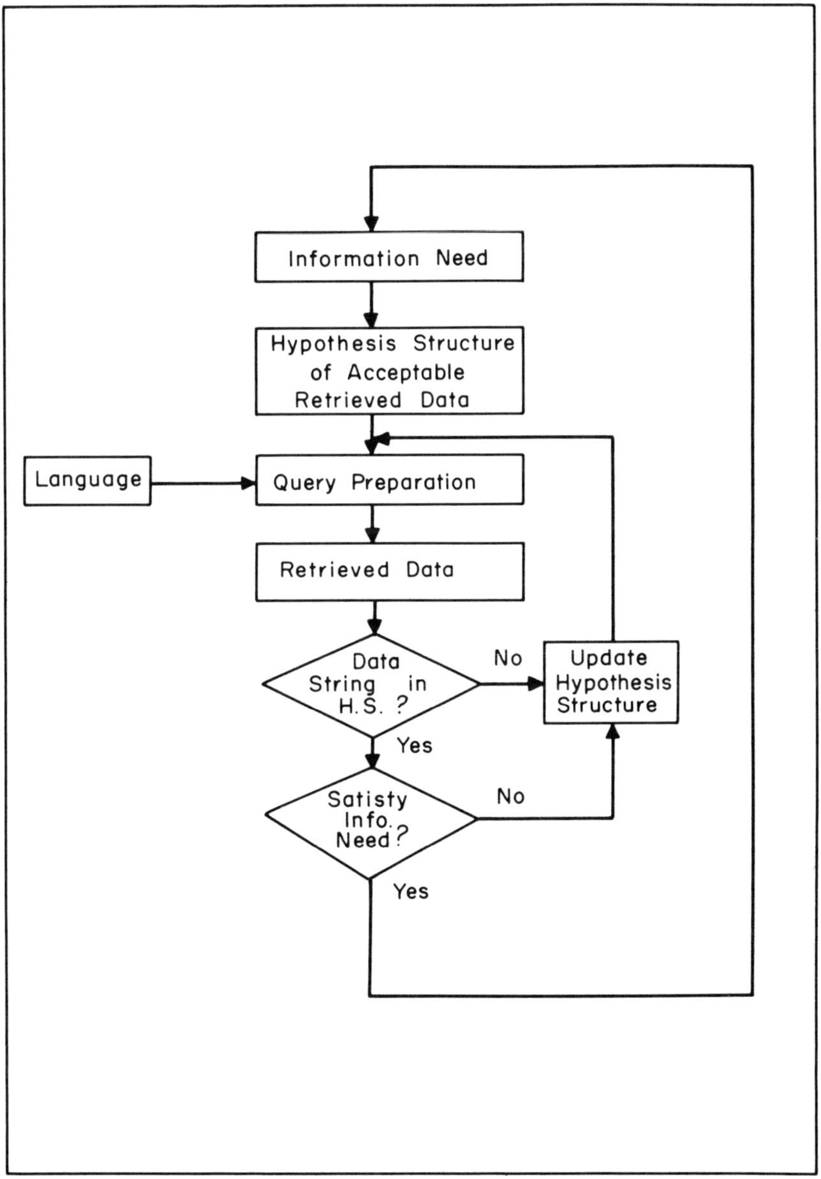

Figure 30–The Cyclic Nature of the Information Retrieval Process

Third, even when the user's information need is satisfied by a system, the value of that system will diminish in direct proportion to the number of iterations required to reach a satisfactory conclusion. The number of iterations a user will permit depends upon his notions of the capabilities of the system, the complexity of the request as he views it, the importance he attaches to the satisfaction of the request, and the time he has for the interaction, regardless of the other factors just mentioned.

Fourth, the accuracy of the document representation is of utmost importance. Errors in document representation can affect the performance of an IS&R system by causing:

(1) failure in the attempt to retrieve a reference that would be retrieved in the absence of the error;

(2) failure in the attempt to retrieve a document given a reference; and

(3) the user to make an incorrect judgment as to the value of a document based upon an inaccurate reference actually retrieved (regardless of his general opinion of the system).

The first two of these effects of error are most likely to be encountered in practice. The types of errors which give rise to these failure conditions can be classified as abstract (conceptual) and concrete (physical). Concrete errors can usually be detected by ordinary editing procedures and include spelling errors, errors in bibliographic citations, format errors, and errors of omission. Abstract errors arise from misinterpretation of the document to be represented, from inadequacies in the mode of representation, or from inconsistencies in the mode of representation. Fugmann calls this type of error *representational distortion* and cites the representational language as the source of such distortion.[10] Concrete errors are troublesome, but they are easily identified and corrected. Abstract errors are not only troublesome but insidious because they very readily slip undetected through editorial machinery and so cause trouble at the most undesirable point, the interface with the user.

Document representation is, from all points of view, an extremely important factor in IS&R system performance and any evaluation of such systems must take full account of this factor.

Search-Request Formulation and Processing

Search-request formulation refers to the process of transforming the user's information need into a form acceptable to the IS&R system. The processing of this formulation results in the identification and retrieval of a set (possibly empty) of documents, usually references, which can then be presented to the user.

Search-request formulation and processing are seldom considered in IS&R system evaluation. Probably this is partly because so little is known about man/machine interaction, but it is no doubt also a result of neglect. Nevertheless, certain facts concerning the user/system interface are clear. For instance, the effort in formulating a search request is inversely proportional to that performed in the processing of the search formulation. Hence, the less work the user has to do to make his search request acceptable to the IS&R system, the better he will like the system. Two examples of the extremes in the ease of formulating a search request are given in Figure 31(a) and (b). The example in Figure 31(a) is typical of formulations required by existing IS&R systems, while that in Figure 31(b) indicates the formulation which the user would like to use and which developers of question-answering systems strive for.[11] Between these extremes one can find systems that perform part of the query formulation automatically.[12] Frequently, however, the user must still contend with some fairly rigid syntactic requirements. Examples of these types of formulations are given in Figure 31(c).

Reasons frequently given for imposing the burden of query formulation on the user can all be subsumed in the observation that automatic query formulation would require too much computer time and would therefore be too costly. Experimental data do not support this observation. It is, in fact, quite possible to operate an IS&R system with automatic query formulation and to do so without financial strain.[13] Most frequently the whole burden of query formulation is imposed upon the user simply because the system was poorly designed or because the designers were lazy. A good IS&R system will enable a user

a.

PROFILE CODING FORM

TERM TYPE	MODE	LOGIC	TERM	TERM WEIGHT
text	N	AND	m. p.	
"	"	OR	mp	
"	S	"	melting point	
"	N	"	melt. pt.	
"	"	"	melting pt.	
name	N	AND	Ethyl benzene	
"	"	OR	Phenyl ethane	
"	"	"	Ethylbenzene	
"	"	"	Benzene, ethyl-	
"	"	"	Et Bz	
"	"	"	Et Ph	
"	"	"	Ph Et	
formula	"	"	$C_2H_5C_6H_5$	

QUESTION WEIGHT ____________

b. What is the melting point of ethylbenzene?

c. • MELTING POINTS *and* CH_3CH_2 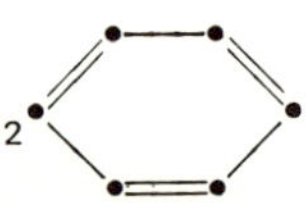

• ETHYLBENZENE, M.P.

Figure 31–Three Forms in Which Search Requests May Be Put to a Retrieval System

(a) A typical search profile; (b) the most desirable form of query; (c) a compromise between (a) and (b)

to express his desires in a manner that is as natural to him as possible. When the user is a human being, the mode of expression illustrated in Figure 31(b) is likely always to be preferred to those of Figures 31(a) and (c).

Formulation and Issuance of Search Responses

The way in which retrieved data is presented to the user by an IS&R system is dependent upon document representation and upon the design of the system given the first dependency. IS&R system output is frequently a source of vexation to the user. Presentation of data in coded form which requires interpretation by the user or which serves only a bookkeeping function will surely merit the user's derision. It is not sufficient that an IS&R system process a user request efficiently and effectively. It is also necessary that the retrieved data be presented in a manner the user can readily understand and evaluate in the context of his information need. To accomplish this goal, the IS&R system must know its users and their habits. While it is not possible to give definite criteria for the proper formulation and presentation of search responses, some general guidelines may be given in the form of factors the user is likely to take into account in judging the value of an IS&R system.

> The data retrieved should be presented to the user in the form requested. Thus, if the user requests a bibliography, he should receive one. On the other hand, if he requests a specific item of data he should receive that item. Of course the ability of a system to satisfy these requirements depends upon the adequacy of its document representation.
>
> The retrieved data should be presented in a form which the user can interpret without recourse to decoding mechanisms. If the user has requested a bibliography, the citations should be presented in common form. If specific data are to be provided, the item(s) should be presented in a way common to the user.
>
> The format in which the data are presented should be clean, well organized, and readily interpretable.

The significance of these factors in evaluating an IS&R system should be obvious.

Cost and Benefit

A person does not, by and large, spend money without expecting something in return. In the case of IS&R systems, what the user buys may not produce any immediate, demonstrable benefit. This is perhaps the most difficult problem an IS&R system must face in its efforts to sell its services. When an IS&R system user does not have to pay directly for the services he receives from the system, as is the case with most libraries, he gives little or no thought to the value received relative to the cash outlay. On the other hand, when the user must pay directly, he is much more sensitive to the benefit derived therefrom. All this suggests that evaluation of IS&R systems in terms of cost/benefit performance is an essential aspect of the overall evaluation of a system.

The problem is to measure benefit in a meaningful way. An approach to such measurement has been suggested by Yovits and Ernst,[14] namely that no benefit is derived from retrieved data except that it produces in the receiver (user) some observable action. Observable actions are themselves hard to measure, but approximations can be obtained. For instance, if, on receipt of one or more document citations, the user decides to go to the library to retrieve the source document, and subsequently does so, this action can be determined and may be considered indicative of the value of the retrieved references. Furthermore, if it can be determined which retrieved references gave rise to the observable action, these can be assumed to be of more value than those retrieved references that did not give rise to any such action.

Of course, it would make life easier if, for example, upon receipt of one or more retrieved references, the user modified his production procedures as a result of which he saved a large amount of money. But so clear-cut an example of benefit derived from the use of an IS&R system seldom presents itself. In any event, the user's judgment of the value of an IS&R system will likely diminish with increasing cost of system use, and the rate of diminution will be affected by the other factors discussed in preceding sections. The reader who is interested in learning more about cost/benefit analyses in IS&R systems should consult King and Bryant.[15]

This chapter has presented a number of factors believed to be central to the development of a successful IS&R system. These factors are therefore important in the overall evaluation of an IS&R system once it has been put into operation. Recently, Fugmann described six postulates for information retrieval that provide a concise summary of most of the points discussed in this chapter.[16] These postulates are:

(1) *Postulate of order:*
The compilation of relevant literature is an order-creating process.

(2) *Postulate of available individual search capacity:*
In searching, disorder can only to a limited extent be compensated for by human patience and efforts of memory.

(3) *Postulate of sufficient degree of order:*
The requirements as to the degree of order rise as the size of the collection and the frequency of searches and the specialization of the inquirer increase.

(4) *Postulate of predictability of the mode of expression of stored concepts:*
Mechanized compilation of relevant literature is based on the predictability of the mode of expression for the stored concepts and for the relations between concepts.

(5) *Postulate of undistorted representation of concepts:*
High order in the mechanical compilation of relevant literature presupposes undistorted representation of concepts.

(6) *Postulate of definability:*
Only search conditions which are definable can be delegated.

These postulates serve as the basis for further research in IS&R system evaluation.

NOTES

1. P. A. Richmond, "The Final Report of the Comparative Systems Laboratory: A Review," *Journal of the American Society for Information Science* 21, no. 2 (1970): 160-162.

2. F. W. Lancaster and J. Mills, "Testing Indexes and Index Language Devices: The ASLIB-Cranfield Project," *American Documentation* 15, no. 1 (1964): 4-13; and Arthur D. Little, Inc., "Centralization and Documentation," *Final Report to the National Science Foundation,* 1963.

3. M. Taube, "A Note on the Pseudo-Mathematics of Relevance," *American Documentation* 16, no. 2 (1965): 69-72.

4. W. J. Paisley and E. B. Parker, "Information Retrieval as a Receiver-Controlled Communication System," in L. Heilprin (ed.), *Education for Information Science* (Washington, D.C.: Spartan Books, 1965), pp. 23-30.

5. F. G. Kilgour, "A Regional Network–Ohio College Library Center," *Datamation* 16, no. 2 (1970): 87-89. See also "Annual Report of the Mechanized Information Center, February 1, 1971 through January 31, 1972." Report of the Ohio State University Libraries to the National Science Foundation (Grant GN-27458), 1972.

6. P. L. Long, K. B. L. Rastogi, J. E. Rush, and J. A. Wyckoff, "Large On-Line Files of Bibliographic Data," *IFIP Congress 71,* August 23-28, 1971, Ljubljana, Yugoslavia.

7. B. C. Landry, *A Theory of Indexing: Indexing Theory as a Model for Information Storage and Retrieval,* Ph.D. Dissertation, The Ohio State University, December 1971, Chapter 5.

8. J. R. Carbonell, "On Man-Machine Interaction: A Model and some Related Issues," *IEEE Transactions on Systems Science and Cybernetics* SSC-5, no. 1 (1969): 16.

9. Landry, *A Theory of Indexing,* p. 207.

10. R. Fugmann, "The Theoretical Foundation of the IDC-System–Six Postulates for Information Retrieval," *ASLIB Proceedings,* February 1972.

11. R. F. Simmons, "Natural Language Question-Answering Systems: 1969," *Communications of the ACM* 13, no. 1 (1970): 15-30.

12. D. A. Thompson, "Interface Design for an Interactive Information Retrieval System: A Literature Survey and a Research System Description," *Journal of the American Society for Information Science* 22, no. 6 (1971): 361-373.

13. Ibid., and R. I. Rubinstein and Arlene Qazi, "Alternatives to Searching Surrogates of Chemical Structures," *Journal of Chemical Documentation* 11, no. 2 (1971): 110.

14. M. C. Yovits and R. L. Ernst, "Generalized Information Systems: Consequences for Information Transfer," in H. B. Pepinsky (ed.), *People and Information* (Elmsford, N.Y.: Pergamon Press, 1970), pp. 1-31.

15. D. W. King and E. C. Bryant, *The Evaluation of Information Services and Products* (Washington, D.C.: Information Resources Press, 1971).

16. Fugmann, "The Theoretical Foundation of the IDC-System."

7

Chemical Structure Representation: Nomenclature

Man's earliest attempts at documentation were made using symbols called pictographs, which are simply drawings of whatever object is being recorded. Thus, an antelope would be represented by a picture of an antelope. Later, as man began coping with abstractions, he started using symbols known as ideographs or ideograms, which represent some attribute or characteristic of the object being depicted. For example, a drawing of an antelope might be used to denote the concept of speed or grace. Ideographs are not primitive. In fact, they form the basis for modern Oriental calligraphy, although the drawings themselves have become quite stylized after millennia of use. In the West, symbols evolved to the point that each stood for a certain sound, establishing the basis for our present phonetic alphabet.

Strings of these symbols, of course, are used in the naming of chemical compounds. If there is sufficient information about the compound, then numerals and punctuation marks may also be used to indicate positions or numbers of components and to serve as delimiters, respectively. If the name completely describes the structure of the compound (as represented by a pictograph), then it is called a systematic name. So-called "trivial" or unsystematic names are those which do not reflect the structure in any way. Most of the chemical names in use fall in between these extremes and are called either semisystematic or semitrivial.[1] Examples in this last category would be 'methane' and 'butene.' One can infer that the former is aliphatic and the latter is olefinic from the suffixes; however, the roots of these terms do not indicate the length of the chain directly, and this connection must be memorized at some point. Moreover, names such as 'butene' do not distinguish between or among the various possible isomers.

Prior to the investigations of Lavoisier in the 1770s, the elemental composition of compounds could only be surmised; systematic nomenclature was therefore out of the question. However, with the quantitative analytical techniques developed by Lavoisier during his work on combustion, it became possible to determine the ratio of the various elements of which a substance was composed. Thus, for example, it was found that carbon dioxide contained carbon and oxygen in the ratio 1:2, and the molecular formula CO_2 was used to denote this substance. Similarly, many other compounds, both organic and inorganic, came to be represented by such formulas. These molecular or empirical formulas can be thought of as the first example of linear notation in chemistry, and they were a necessary precursor to systematic chemical names.

The step from such line-formula notations to structural formulas did not occur until about 1850. When one considers the formidable nature of the task of determining how atoms were interrelated in such compounds as morphine, quinine, and strychnine, all of which were known by the beginning of the nineteenth century, it is not surprising that Wöhler wrote the following in a letter to Berzelius in 1835:

> Organic chemistry just now is enough to drive one mad. It gives one an impression of a primeval tropical forest, full of the most remarkable things, a monstrous and boundless thicket, with no way to escape, into which one may well dread to enter.

The problem faced by early chemists was compounded by the discovery by Liebig in 1873 that silver fulminate has the same composition as silver cyanate, although the two substances have radically different properties. Berzelius subsequently found that tartaric acid and racemic acid have the same composition, $C_4H_6O_6$, and he introduced the term 'isomerism' to name the phenomenon. Following this came the discovery of *optical* isomerism through the work of Nicol, Malus, Biot, Pasteur, and many others.[2]

The remarkable fact about these discoveries is that the chemists involved did not have a very precise conception of the structures of the substances they were working with until 1858, when Kekule and Couper independently advanced the theory of structure, which now usually bears Kekule's name. (Kekule's theory appeared first because of a delay in the publication of Couper's work. For a more detailed account see Leicester and Klickstein.[3]) Kekule based his theory on the premise that carbon has the same valence (four) in complicated organic compounds as it exhibits in simple compounds such as methane, carbon tetrachloride, and carbon dioxide. He introduced a graphical representation in which available valences were shown as short lines emanating from the appropriate atomic symbols:

From these ideas came the Kekule formulas for compounds, such as

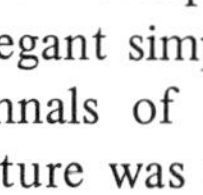

methane

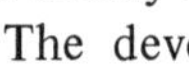

carbon dioxide

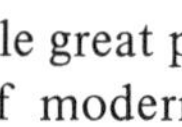

ammonia

and the more complicated ones as well.

The elegant simplicity of his theory merits Kekule great prominence in the annals of chemistry. The development of modern chemical nomenclature was made possible by his work. It will be shown in the next two chapters that the other principal means of chemical structure representation also have their basis in the Kekule theory of structures.

THE NEED FOR VOCABULARY CONTROL

In everyday life, chemists may use any of a variety of names for a compound. A trivial name may be used for the sake of brevity, a systematic name may be chosen to elucidate a chemical structure, and a semisystematic name can be used as a satisfactory compromise. In the laboratory, labels such as Compound No. 9 and Steroid 17 are common, particularly in cases where the structure is in doubt. Systematic names may be used throughout a journal article for the sake of rigor and consistency of style, but trivial or semisystematic names may be used in a lecture or discussion for greater fluency and ease of pronunciation.

This is as it should be. However, as was noted in Chapter 3, with the exception of certain specialized indexes which are used primarily for current awareness, much can be gained by standardizing the terms used for indexing purposes. Vocabulary control, the systematic selection of preferred index terms, is a necessary part of subject indexing if the user of the index is to know the location of the information he is seeking. If no such control is exercised, scattering of information will result; that is, information will appear under synonymous headings which are scattered throughout the alphabet. The difficulties that this situation can create are, of course, greatly magnified if the indexes are to be cumulated and used for retrospective searching.

All scientific fields have nomenclature difficulties, but the problem in chemistry is exacerbated by the tremendous number of known compounds and the rate at which new ones are being produced. Estimates of the total number of known compounds range from three to four million.[4]

Tradition vs. Standardization

It has long been recognized that standardization of chemical nomenclature is highly desirable, and because of the special problems arising from the need to translate foreign scientific works, there has been substantial international cooperation in achieving this standardization. A concise but highly informative history of this cooperation, complete with a sixty-three-item bibliography, has been prepared by Loening.[5]

Currently, the International Union of Pure and Applied Chemistry (IUPAC) and the International Union of Biochemistry (IUB), through their nomenclature commissions, are responsible for the continued development of definitive and internationally acceptable chemical nomenclature. References to the publications of these international organizations can be found in Loening's bibliography and in the American Chemical Society's *Handbook for Authors.*[6] Although it is by now somewhat dated, perhaps the most convenient source for the average reader would be the *Handbook for Chemical Society Authors,* which contains reprints of the definitive rules for nomenclature prior to 1961.[7] In addition, there has been an announcement that IUPAC has completed a second edition of its *Nomenclature of Inorganic Chemistry,* a third edition of *Nomenclature of Organic Chemistry,* and a new edition of *Manual of Symbols and Terminology for Physicochemical Quantities.*[8]

In spite of the availability of such definitive rules, they may or may not be used by indexers to effect vocabulary control. There are basically three reasons for this. First, many trivial names have become so well established during the history of chemistry that they may always be used in preference to a systematic name. Second, the definitive systematic name may violate sound indexing principles, thus making its adoption a disservice to the user of the index. Third, in those cases involving a choice between or among trivial, semisystematic, and systematic names, all of which are currently in widespread use, a preference may be shown to the term thought to be most widely used since a user will probably look for it under that name first. Moreover, to say that rules are definitive is not to say that they are comprehensive, and compounds keep appearing for which no adequate rules exist.

It is for such reasons that indexing services often employ their own nomenclature systems for selecting preferred index terms. The best example is probably that of the Chemical Abstracts Service, and the handiest reference work which outlines this system continues to be *The Naming and Indexing of Chemical Compounds from Chemical Abstracts.*[9] The discussion which follows focuses on just a few of the basic principles that underlie the rules of this system.

Inverted Names

One of the types of main headings mentioned in Chapter 3 was the inverted form, and the following three entries were used for illustration:

Acetic acid, trichloro-
Light, infrared
Tin, analysis

Inversion is one of the techniques employed to minimize the scattering of information in an index. In the case of chemical compound names, a portion of the entire compound is selected on the basis of its indexing value and is named as a "parent" compound. All remaining portions of the entire compound are then named as substituents of the parent following the comma of inversion. As a rule, if there are several candidates for the parent name, the largest one is chosen. However, this rule is sometimes overridden in the interest of treating like groups alike.

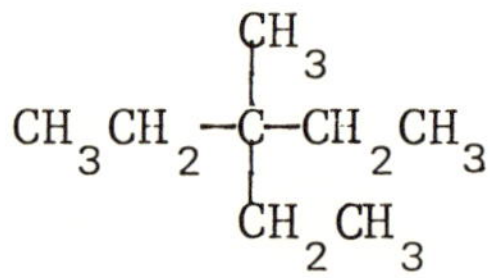

Pentane, 3-ethyl-3-methyl-

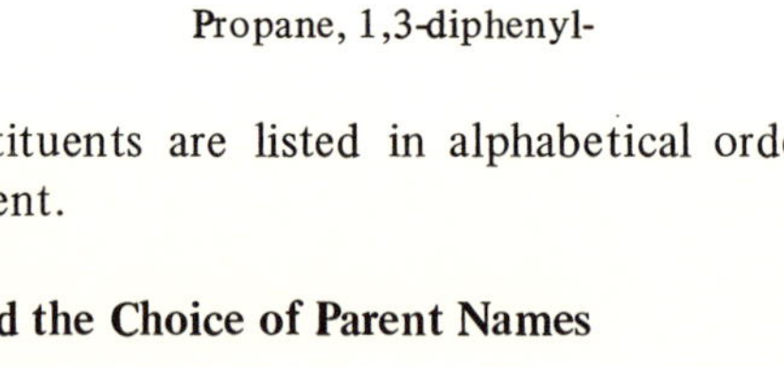

Propane, 1,3-diphenyl-

Note that substituents are listed in alphabetical order following the name of the parent.

Unsaturation and the Choice of Parent Names

When choosing the parent of an alkene, one looks for the longest continuous chain containing double bonds. If two chains are of the same length and both contain double bonds, then the one with the greater number of double bonds is chosen as parent.

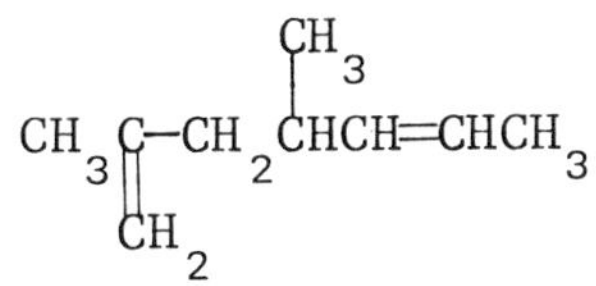

1,5-Heptadiene, 2,4-dimethyl-

Note that the principle of low numbering is employed. This principle may be stated formally as follows: of two series of numbers (locants), that series is lower which contains the lowest number (locant) at the first point of difference. Thus, 1, 2, 3, 5, 7 is lower than 1, 2, 4, 5, 6.

Simple alkynes are named in a manner completely analogous to the corresponding alkenes (with the exception, of course, of trivial names which are retained for indexing purposes). In choosing the parent of an alkyne which contains double bonds and is branched, one looks first for the chain containing the greatest number of carbon atoms. If two chains in such a compound are of equal length, the one is chosen as parent which has the greater number of double bonds.

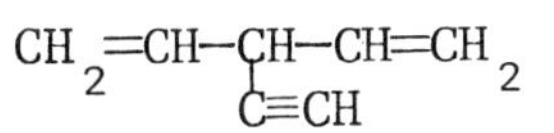

1,4-Pentadiene, 3-ethynyl-

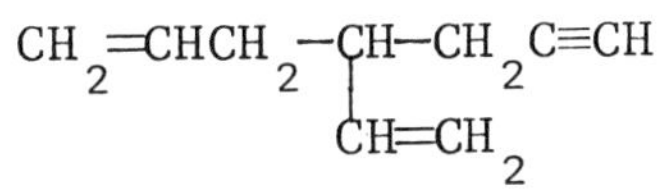

1-Hepten-6-yne, 4-vinyl-

Simple Cyclic Hydrocarbons

Cyclic hydrocarbons are usually named in a manner analogous to their acyclic counterparts. The choice of parent compound names is determined by the same principles of selecting the largest parent and treating like groups alike.

If a straight-chain alkane is attached directly to a cycloalkane having either fewer or the same number of carbon atoms in its ring, then the compound is named as a cycloalkyl derivative of the straight-chain alkane.

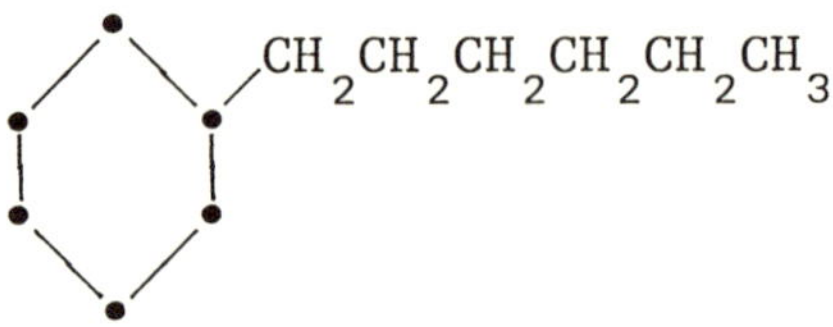

Hexane, 1-cyclohexyl-

When cycloalkanes are directly united, they are named as cycloalkyl derivatives of the larger ring involved in the union.

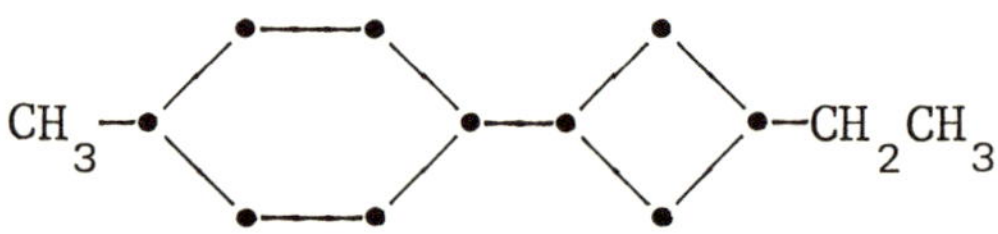

Cyclohexane, 1-(3-ethylcyclobutyl)-4-methyl-

If both rings are the same size, then the compound is named by attaching the prefix 'bi-' to the appropriate cycloalkyl name. Both rings are numbered from the point of attachment, and 'primed' numbers are used as locants for one of the rings. Locants not having primes are given low numbers wherever possible.

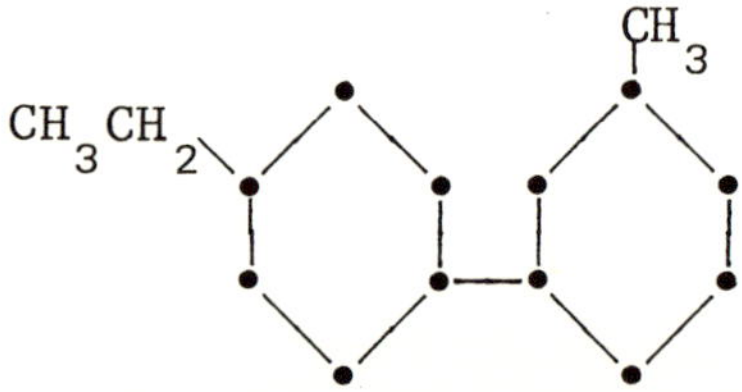

Bicyclohexyl, 4′-ethyl-3-methyl-

When benzene is directly attached to another cyclic hydrocarbon, the larger ring is chosen as the parent. If benzene is attached to a smaller cyclic hydrocarbon more than once, then the smaller ring is chosen in accordance with the principle of treating like groups alike.

When it is attached to cyclohexane, cyclohexene, or cyclohexadiene, benzene is chosen as the parent. In the special case where two benzene rings are directly united, the name 'biphenyl' is used, and one ring receives primed locants for its substituents if needed.

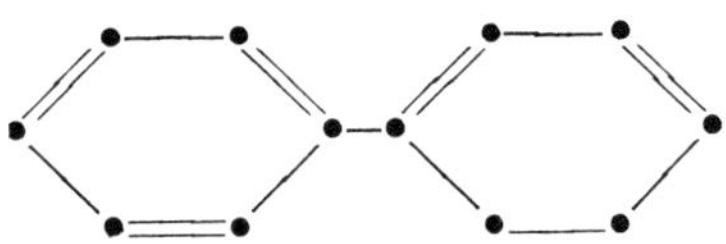

Benzene, (1,3-cyclohexadien-l-yl)-

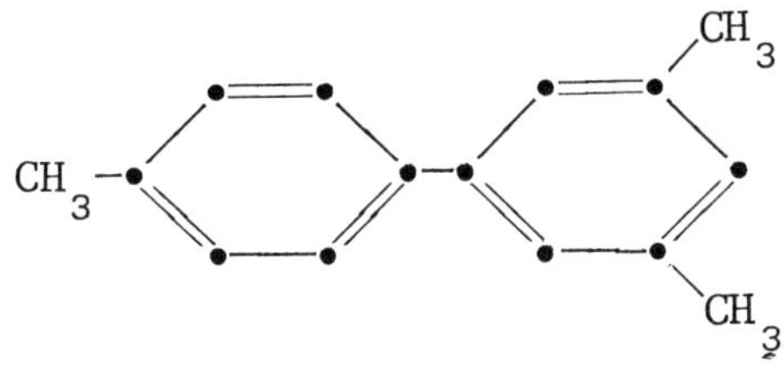

Biphenyl, 3,4′,5-trimethyl-

Because of their size and importance, the other aromatic hydrocarbons are almost always chosen as the parent when substituted by alkyl, alkenyl, and/or alkynyl groups.

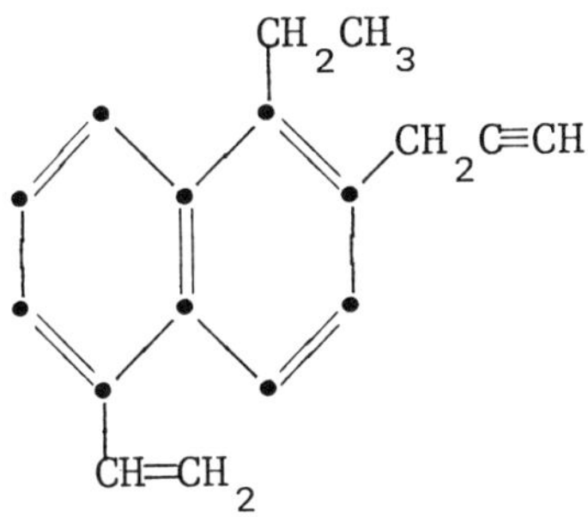

Naphthalene, 1-ethyl-2-(2-propynyl)-5-vinyl-

Simple Functional Substituents

Many simple substituted compounds are named by selecting the appropriate hydrocarbon as parent and listing the substituents following the comma of inversion.

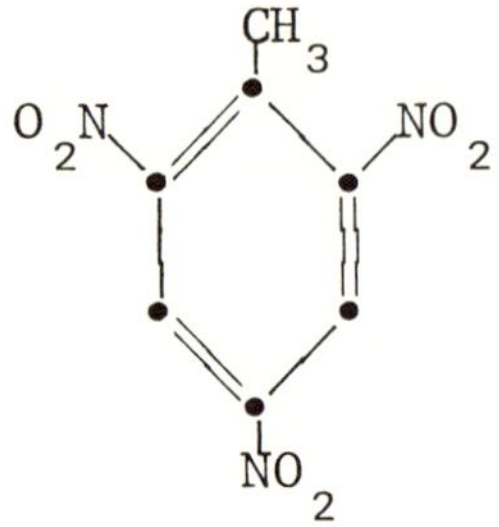

Toluene, 2,4,6-trinitro-

Benzene, *p*-dichloro-

Note that the traditional positional indicators *ortho-*, *meta-*, and *para-* are used for simple disubstituted benzenes.

Names of the more reactive functional groups are normally incorporated in the name of the parent either explicitly or implicitly. To exemplify the procedure, the naming of carboxylic acids will be discussed. This particular class of compounds has been chosen because the names of acyl halides, aldehydes, amides, imides, nitriles, and certain other functions are almost always closely related to the names of the acids to which they correspond.

The simplest series of carboxylic acids is the series R-COOH, where R represents either a hydrogen atom or an unbranched alkyl group. The first five members of this series have trivial names which are retained. All others are named by replacing the final 'e' with the suffix 'oic acid.' The first eight members of this series are as follows:

Formic acid	H—COOH
Acetic acid	CH_3—COOH
Propionic acid	CH_3CH_2—COOH
Butyric acid	$CH_3CH_2CH_2$—COOH
Valeric acid	$CH_3CH_2CH_2CH_2$—COOH
Hexanoic acid	$CH_3CH_2CH_2CH_2CH_2$—COOH
Heptanoic acid	$CH_3CH_2CH_2CH_2CH_2CH_2$—COOH
Octanoic acid	$CH_3CH_2CH_2CH_2CH_2CH_2CH_2$—COOH

Numbering starts with the carboxyl group and proceeds back through the longest continuous chain in the structure if branching occurs.

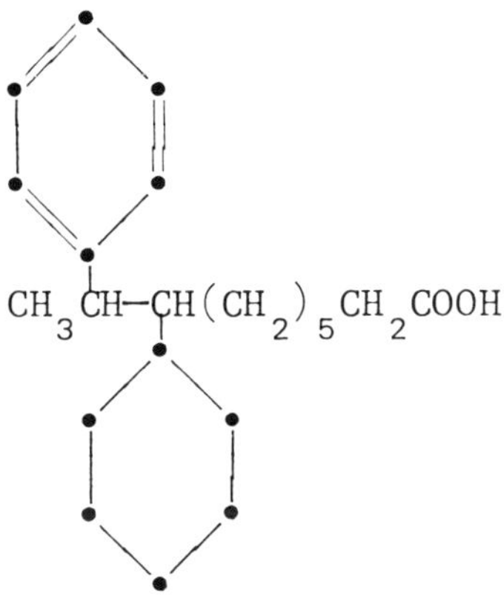

Decanoic acid, 8-cyclohexyl-9-phenyl-

In the series of acids with two carboxyl groups attached either to each other or to the terminal carbon atoms of a straight-chain alkane, the first four members have trivial names which are retained. All other members of this series are named by adding 'dioic acid' to the name of the straight-chain alkane which has the same number of carbon atoms as the acid. Note that *CA* retains the name 'adipic acid' when that compound is unsubstituted.

Oxalic acid	HOOC–COOH
Malonic acid	$HOOC{-}CH_2{-}COOH$
Succinic acid	$HOOC{-}CH_2CH_2{-}COOH$
Glutaric acid	$HOOC{-}CH_2CH_2CH_2{-}COOH$
Adipic acid	$HOOC{-}CH_2CH_2CH_2CH_2{-}COOH$

These compounds are numbered from one carboxyl group to the other in the direction which observes the principle of low numbering.

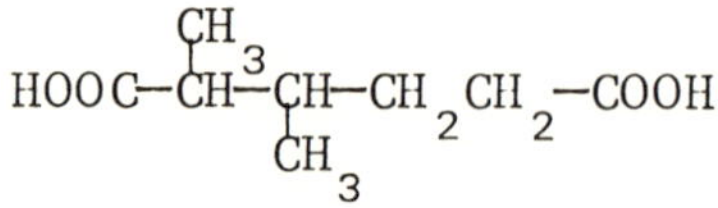

Hexanedioic acid, 2,3-dimethyl-

When a carboxyl group is attached to a hydrocarbon chain which contains double and/or triple bonds, the parent acid is chosen by starting with the carboxyl group and numbering back through the longest continuous chain even if it contains no double or triple bonds. If the longest chain is unsaturated, then the acid is given names such as 'alkenoic acid' and 'alkadienynoic acid.' Whenever there is a choice between two chains, the one with the greater number of double and triple bonds is selected. Double bonds are given preference over triple bonds if all other factors are equal.

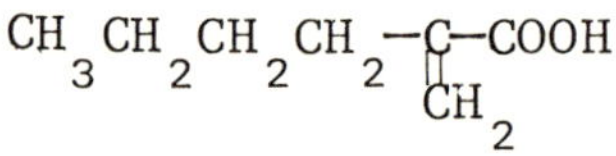

Hexanoic acid, 2-methylene-

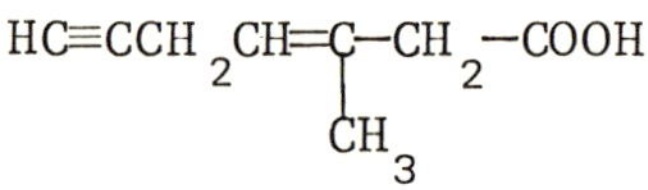

3-Hepten-6-ynoic acid, 3-methyl-

When two carboxyl groups are attached to an unsaturated hydrocarbon chain, the resulting acid is assigned names such as 'alkenedioic acid,' 'alkadiynedioic acid' and 'alkenynedioic acid,' in a manner analogous to the class of alkanedioic acids.

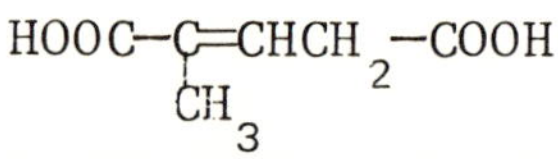

Pentenedioic acid, 2-methyl-

2-Hepten-5-ynedioic acid

When three or more carboxyl groups are attached to a hydrocarbon chain, the parent acid is named by adding 'tricarboxylic acid,' 'tetracarboxylic acid,' and so forth, to the name of the longest continuous chain to which all the carboxyl groups are attached. The locants designating the positions of the carboxyl groups should precede the name of the parent unless double or triple bonds are present in the longest chain. When this occurs, the locants for the double or triple bonds are put in their usual places, and the locants for the carboxyl groups are placed immediately before the 'carboxylic acid' part of the name.

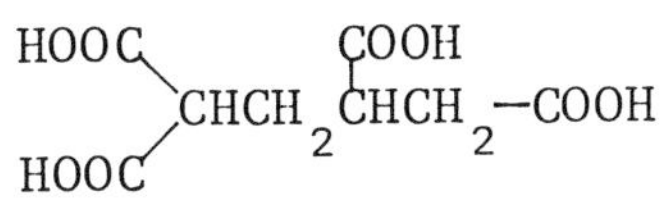

1,1,3,4-Butanetetracarboxylic acid

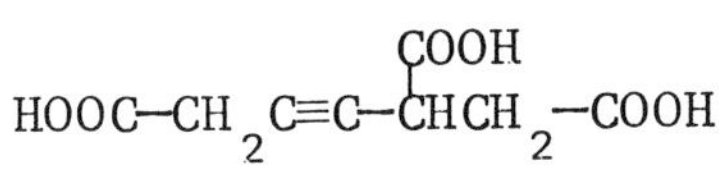

3-Pentyne-1,2,5-tricarboxylic acid

"Carboxylic acid" nomenclature is also used for acids having one or more carboxyl groups attached directly to nonaromatic cyclic hydrocarbons.

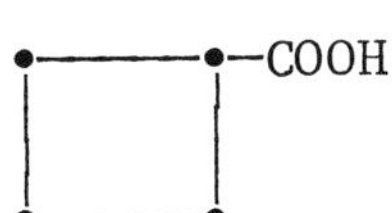

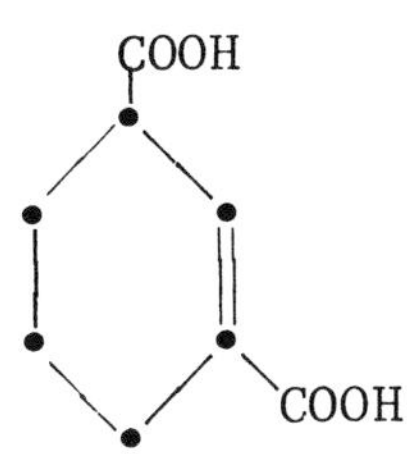

Cyclobutanecarboxylic acid   1-Cyclohexene-1,3-dicarboxylic acid

CA retains many trivial names for aromatic carboxylic acids. Examples are benzoic, toluic, naphthoic, phthalic, isophthalic, and terephthalic acids. However, when two or more carboxyl groups are attached to naphthalene, and when three or more are attached to benzene, the resulting acids are given "carboxylic acid" names.

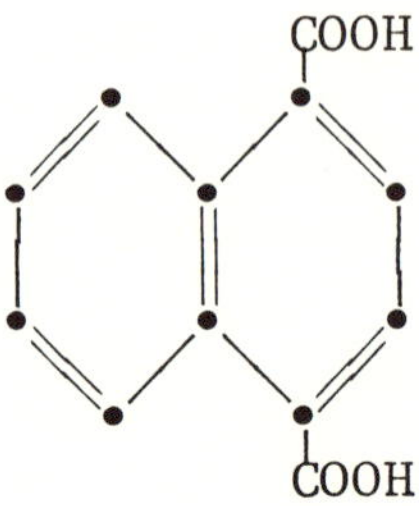

1,4-Naphthalenedicarboxylic acid

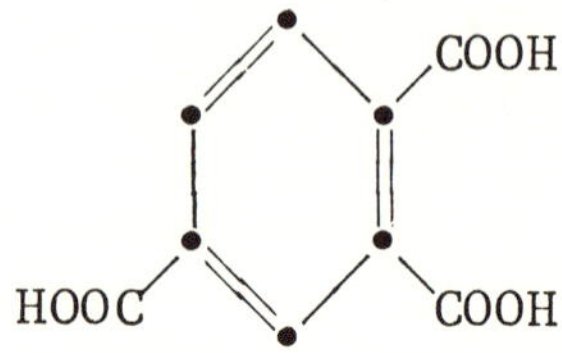

1,2,4-Benzenetricarboxylic acid

It was mentioned earlier that several other classes of compounds followed the naming of carboxylic acids rather closely. Examples of various names and structures are shown below to illustrate this:

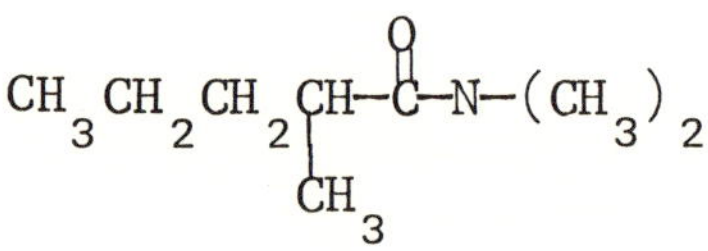

Valeramide, *N,N,*2-trimethyl-

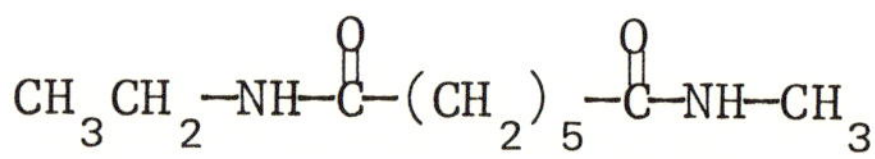

Heptanediamide, *N*-ethyl-*N'*-methyl-

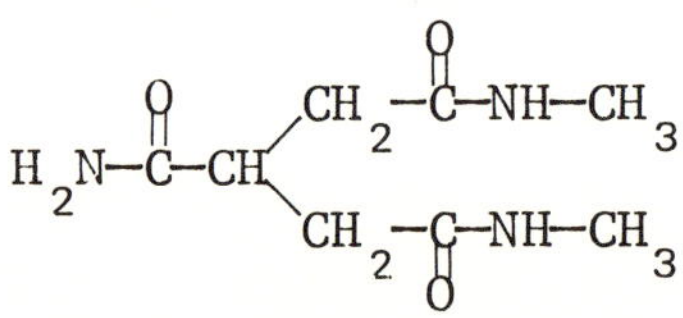

1,2,3-Propanetriamide, N^1,N^3-dimethyl-

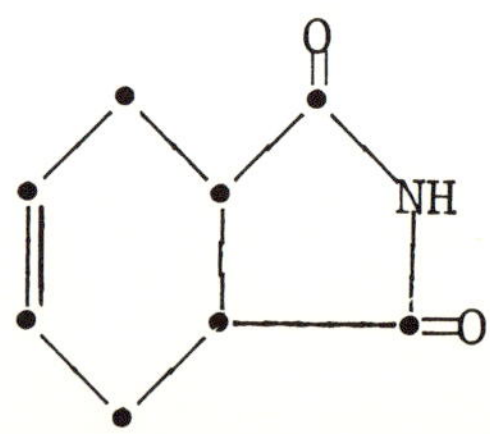

4-Cyclohexene-1,2-dicarboximide

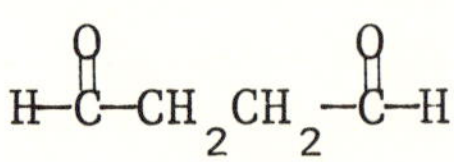

Succinaldehyde

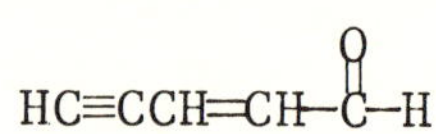

2-Penten-4-ynal

$N{\equiv}C{-}CH_2CH_2CH_2CH_2CH_2CH_2CH_2{-}C{\equiv}N$

Nonanedinitrile

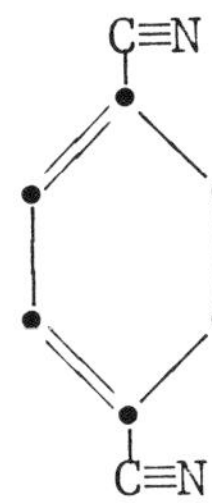

Terephthalonitrile

$CH_2{=}CHCH_2CH_2{-}C{\equiv}N$

4-Pentenenitrile

Unfortunately, names are not always assigned in such a regular fashion, and after a user has attempted to find specific entries for compounds such as the various anilides, toluidides, and xylidides, he will realize that the process often requires both a certain amount of imagination and the judicious use of *CA*'s notes and cross references.

Several of the basic rules for naming compounds apply generally, regardless of the function involved. Examples are: picking the largest parent, treating like groups alike, and applying the principle of low numbering. Such rules are equally applicable to compounds whose names are not connected in any immediate way with the naming of carboxylic acids. There are, however, specific problems which may result in names that seem strange or counterintuitive. Illustrative examples have been chosen for the following classes of compounds: ketones, alcohols and phenols, amines, and ethers.

Whenever a cyclic hydrocarbon other than benzene or naphthalene is attached to one side of a carbonyl group and a hydrocarbon chain is attached to the other, the hydrocarbon chain is chosen as the parent and the cyclic hydrocarbon is named as a substituent on the "1" position.

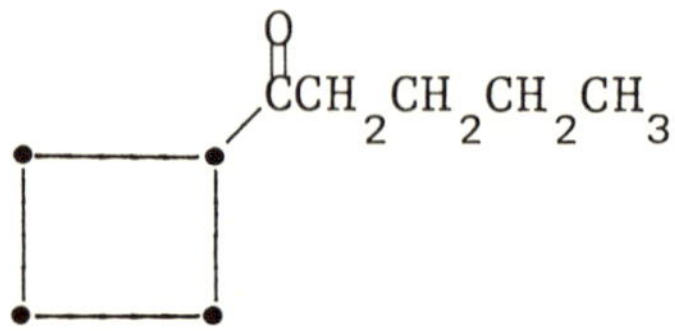

1-Pentanone, 1-cyclobutyl-

An exception is the case in which a methyl group is opposite a cyclic hydrocarbon other than benzene or naphthalene. Compounds of this type are named using the class name 'ketone.' The name 'ethanone' is not used.

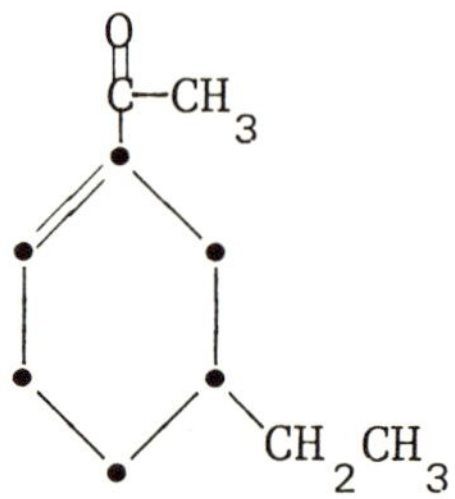

Ketone, 5-ethyl-1-cyclohexen-1-yl methyl

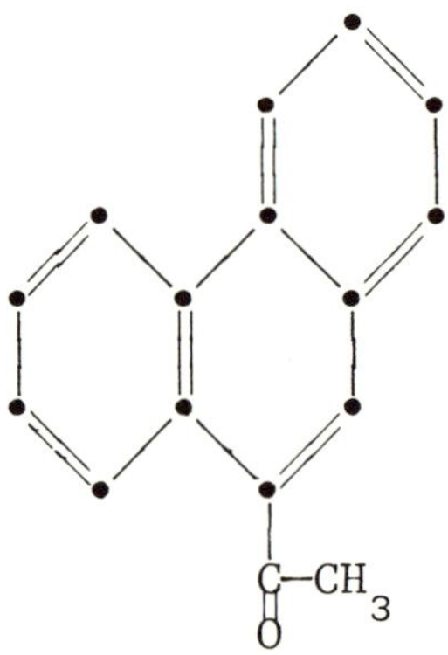

Ketone, methyl 9-phenanthryl

Hyphens are not used following either of the groups in the inverted part of the name because 'ketone' is a class name and not the name of a parent compound. In addition, because of the spacing which results from this practice, fewer parentheses are needed to provide clarity.

With the exception of benzophenone, all ketones with cyclic hydrocarbons attached to both sides of a carbonyl group are named using the class name 'ketone.'

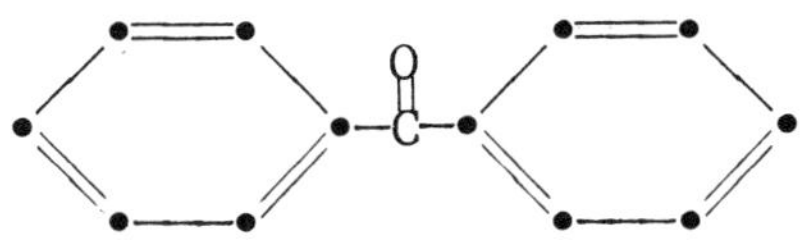

Benzophenone

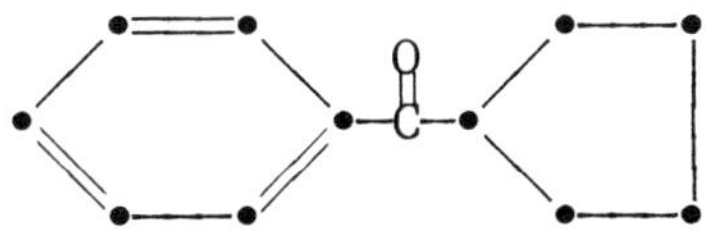

Ketone, cyclopentyl phenyl

If benzene or naphthalene is attached to one side of a carbonyl group and a hydrocarbon chain is attached to the other, the resulting structure may be thought of as deriving from the union of an acyl group and the aromatic compound. These structures are named by adding the suffixes 'phenone' or 'naphthone' to the roots of the corresponding acids plus a connective 'o.' The locants for the acyl groups are the same as for their corresponding acids, and primed numbers are used as locants for the benzene and naphthalene rings.

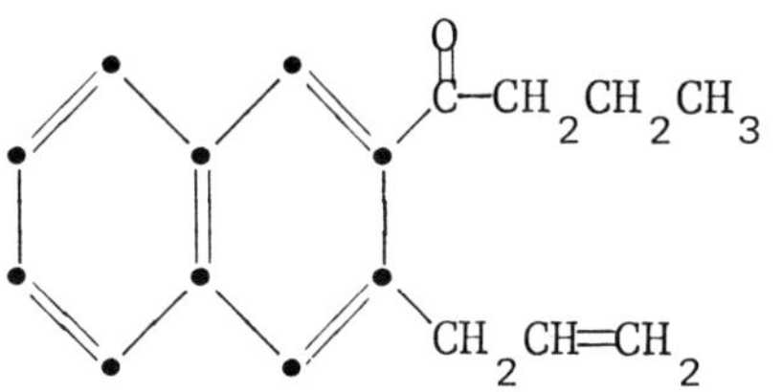

2′-Butyronaphthone, 3′-allyl-

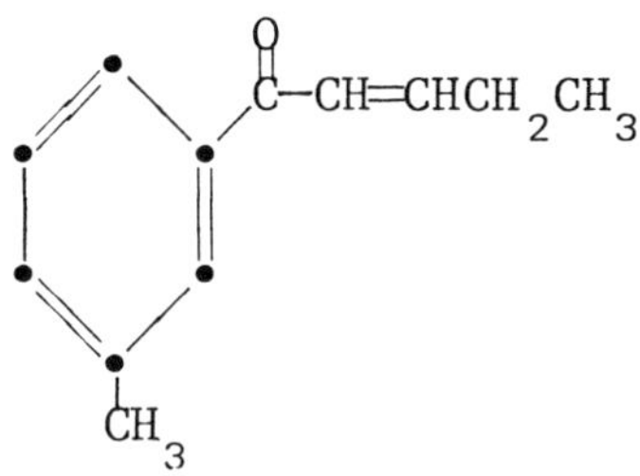

2-Pentenophenone, 3′-methyl-

The naming of alcohols is fairly regular. However, this class of compounds shares an interesting problem with aldehydes and ketones: keto-enol tautomerism. *CA*'s policy, as stated in *The Naming and Indexing of Chemical Compounds,* is that "a tautomeric compound is usually indexed as an aldehyde or ketone as the more stable form," but it is named as an alcohol or phenol "if the enol form is individually mentioned."[10] Thus, to be sure he has found everything of potential interest to him, a user must examine the headings for all the possible tautomeric forms of a compound. Nor is the problem solved by examining the formula index, since derivatives of the compounds will be filed at different formulas. The problem is exacerbated by the continued use of trivial names which further obscure tautomeric relationships. For example, malonaldehydic acid and 3-hydroxyacrylic acid, from the standpoint of a person interested in keto-enol tautomers, would be thought of as derivatives of acetaldehyde and vinyl alcohol; however,

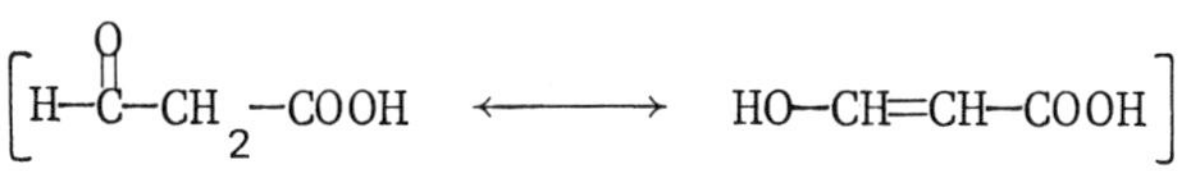

Malonaldehydic acid (Acetaldehyde, carboxy-) | Acrylic acid, 3-hydroxy- (Vinyl alcohol, 2-carboxy-)

no entries would appear at these headings because *CA*'s rules forbid it. Further discussion of the storage and retrieval of tautomeric information can be found in a paper by Davis.[11]

The naming of phenols is not as regular as the naming of alcohols, primarily because of the numerous trivial names which are retained, such as 'resorcinol,' 'hydroquinone,' and 'pyrogallol.'

The retention of trivial names may be the greatest difficulty in naming amines, and it has already been pointed out that this also complicates the naming of various amides. Using names such as 'benzylamine,' 'toluidine,' and 'xylidine' forces the indexer to introduce yet another kind of locant, namely the Greek letter alpha. Whenever alpha appears in a series of locants, it is placed following Roman letters but before numerals.

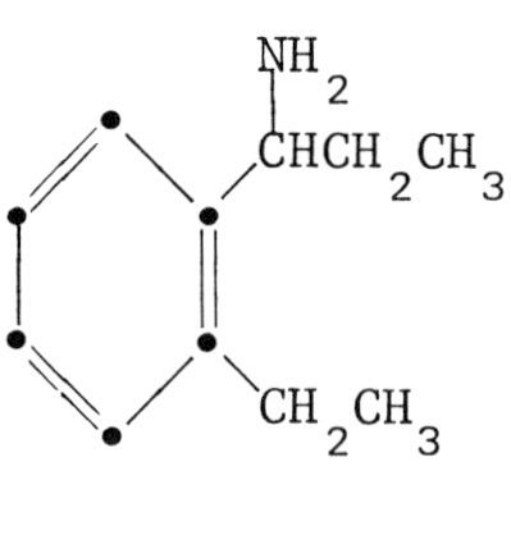

Benzylamine, *o*,α-diethyl-

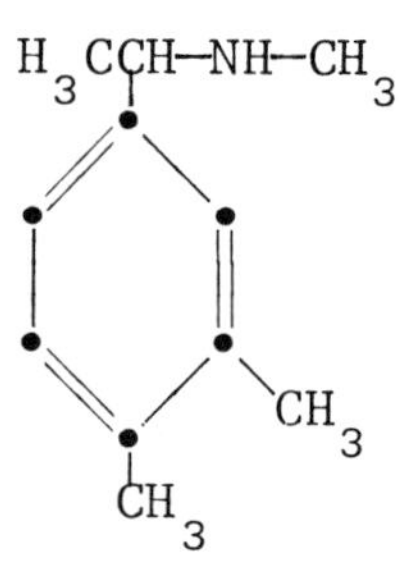

Benzylamine, *N*,α3,4-tetramethyl-

CA uses primes to distinguish between and among *N*-substituted compounds as long as this does not result in ambiguity. Where primes are not adequate, superscript numerals are used instead.

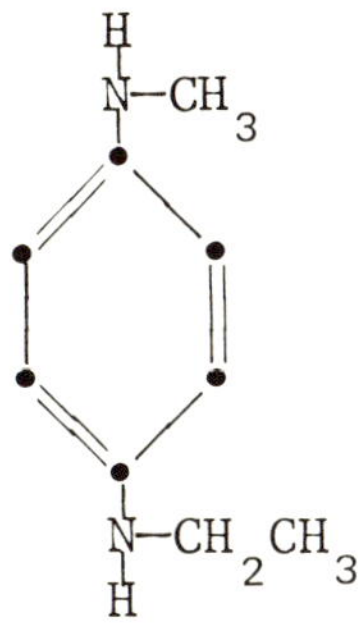

p-Phenylenediamine, *N*-ethyl-*N*′-methyl-

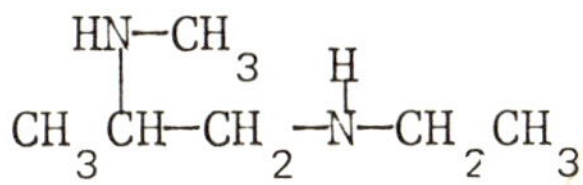

1,2-Propanediamine, N^1-ethyl-N^2-methyl-

Ethers are interesting in that they, like some of the ketones, are often named by using their class name. Once again, since 'ether' is not the name of a parent compound, no hyphens are used following the group or radical names.

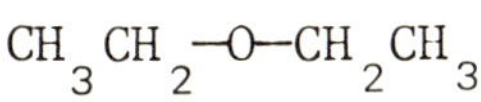

Ether, diethyl

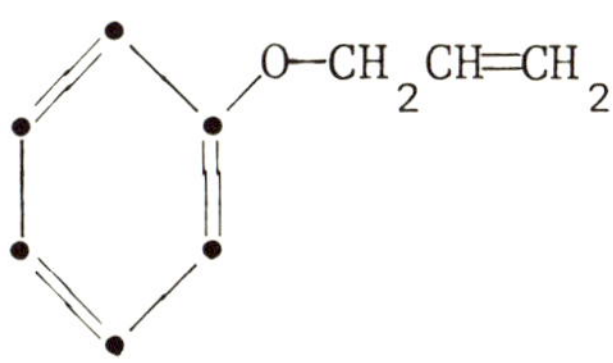

Ether, allyl phenyl

When there is an ether linkage to polycyclic hydrocarbons such as naphthalene and anthracene, the hydrocarbon is chosen as the parent.

The names 'anisole' and 'phenetole' are retained for methyl phenyl ether and ethyl phenyl ether.

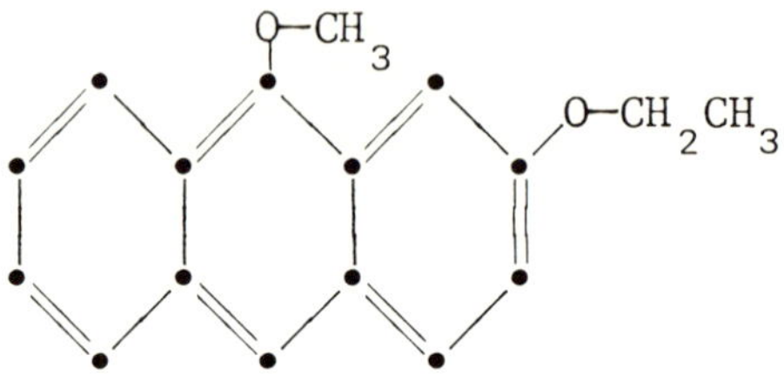

Anthracene, 2-ethoxy-9-methoxy-

In general, when two or more ether linkages occur in the same compound, the largest and most central hydrocarbon is selected as the parent. However, when three or more ether linkages occur in the same compound and the compound is symmetrical about an oxygen atom, then an ether class name is assigned to the structure.

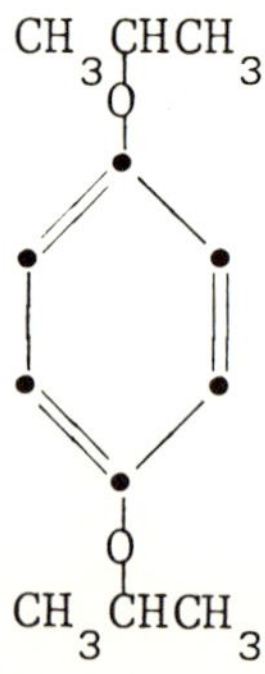

Benzene, *p*-diisopropoxy-

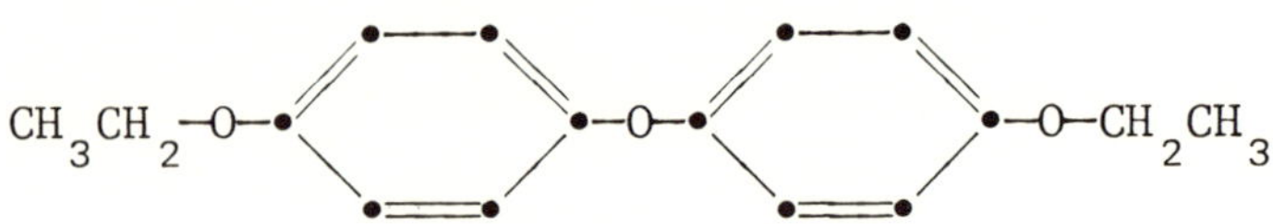

Ether, bis(*p*-ethoxyphenyl)

When two alkoxy groups are attached to the same carbon atom, the resulting compound is normally named as an acetal of the corresponding aldehyde. However, *CA* makes an exception for the compound dimethoxymethane.

Finally, some of the smaller oxygen heterocycles are often named by using "epoxy" nomenclature. The largest hydrocarbon chain is chosen and numerals are used as locants with the group name 'epoxy' to show the points of attachment of the oxygen atom to the chain.

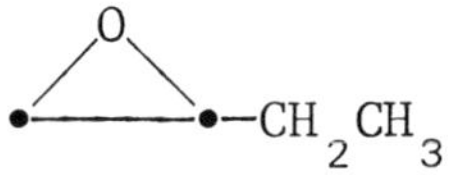

Butane, 1,2-epoxy-

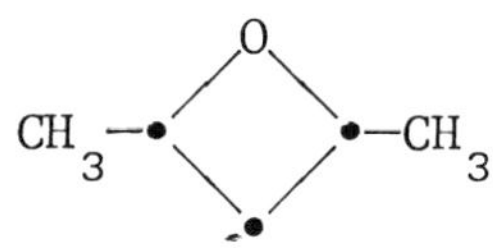

Pentane, 2,4-epoxy-

Conjunctive Names

One type of name which is definitely more common in indexes than in the laboratory is the so-called conjunctive name. Its primary purpose is to maximize the effects of inversion by incorporating as much information as possible into the name of the parent compound. For example, the name 'Naphthaleneacetic acid' would be used instead of 'Acetic acid, naphthyl-' because of the relative importance of naphthalene. Other functions are treated similarly. This practice necessitates the use of Greek letters as locants for the side chains, but the advantages of this practice are believed to outweigh such inconveniences.

Order of Precedence of Functions

In a nomenclature system such as that used by *CA* in its subject indexes, names are dependent on an order of precedence of functions in cases where a compound contains more than one functional group. Thus, an aldehyde that contains a carboxyl group is not named as an aldehyde but rather as an acid. An amine that contains a hydroxyl group is named as an alcohol rather than as an amine, and so forth. By following a series of rules based on such a hierarchy, one arrives at the name of a parent compound, and all other functions and groups are named as substituents. The choice would be arbitrary except that an index of chemical compounds must follow some sort of order of precedence for the sake of consistency. The names and order of precedence of several functions used by *CA* are shown in Table 1. A more complete listing can be found in *The Naming and Indexing of Chemical Compounds.*[12]

There are two types of names for the groups formed from the amide function, depending on whether the group is attached to the parent through the carbon atom or the nitrogen atom.

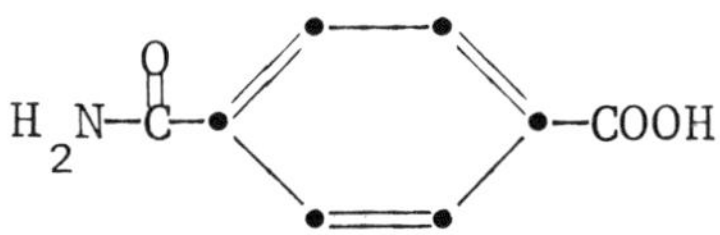

Benzoic acid, *p*-carbamoyl-
(*CA* uses 'Terephthalamic acid')

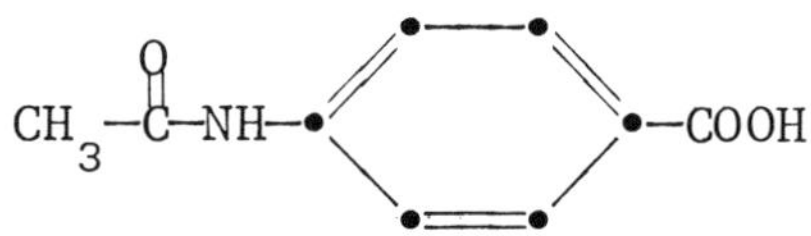

Benzoic acid, *p*-acetamido-

In addition, the group name used to indicate the presence of a ketone function will vary depending on the type of compound and the point of attachment.

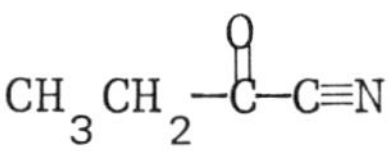

Butyronitrile, 2-oxo-

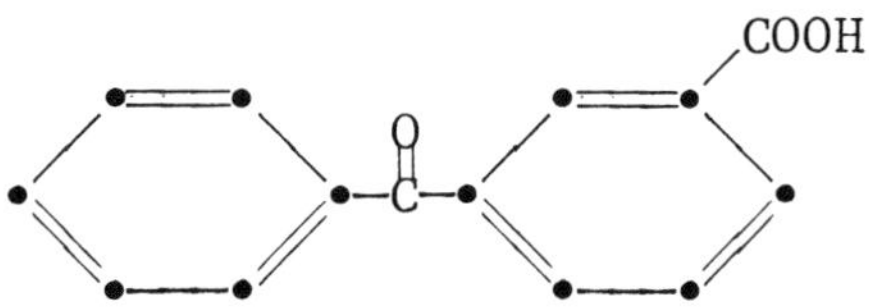

Benzoic acid, *m*-benzoyl-

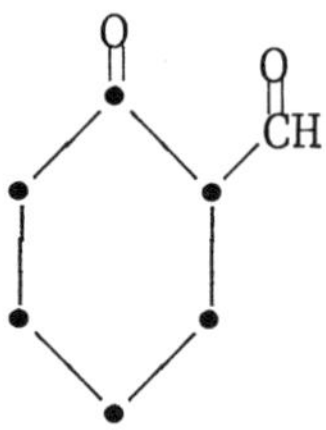

Cyclohexanecarboxaldehyde, 2-oxo-

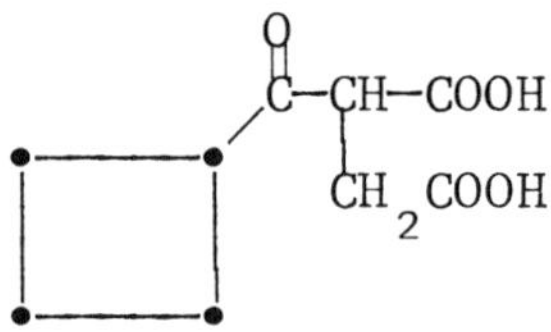

Succinic acid, (cyclobutylcarbonyl)-

Table 1–Partial List of Functional Groups, Their *CA* Names, and Their Precedence (in descending order)

FUNCTION	GROUP	GROUP NAME
Carboxylic acid	HOOC–	carboxy
Carboxamide	$H_2N–C=O$	carbamoyl
	R–C(=O)–NH–	___amido
Carboximide	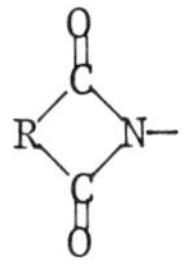	___imido
Carboxaldehyde	O=CH–	formyl
Nitrile	N≡C–	cyano
Ketone	=O	oxo
	R–C(=O)–	___oyl
	R'–C(=O)	___carbonyl
Alcohol	HO–	hydroxy
Phenol	HO–	hydroxy
Amine	$H_2N–$	amino
	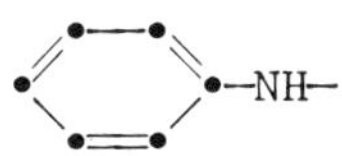	anilino
	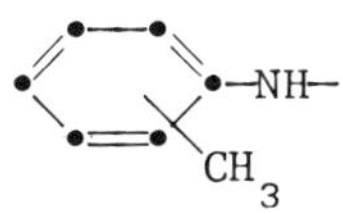	toluidino
	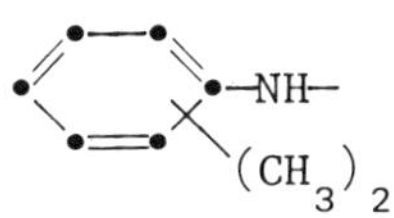	xylidino
Ether	R–O–	___oxy

Curves and brackets are used to provide clarity by helping to show what is attached to what. Their use becomes especially important in naming compounds involving mixed functions. Curves are used around all group names consisting of two parts, and brackets are used to enclose complex group names having three or more parts.

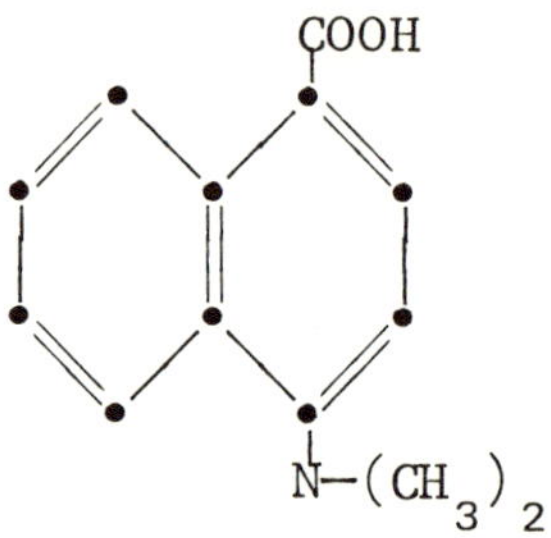

1-Naphthoic acid, 4-(dimethylamino)-

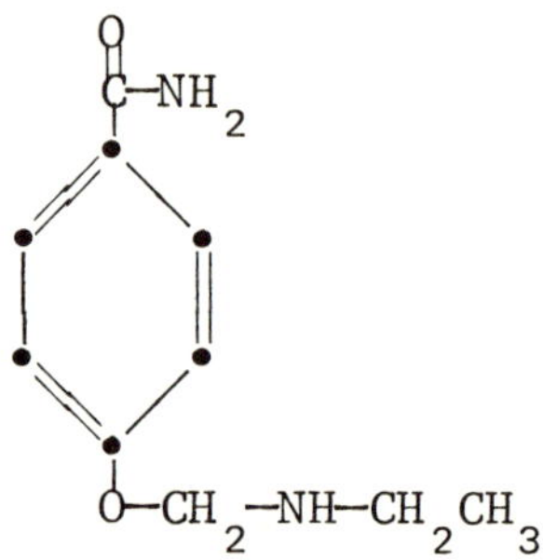

Benzamide, *p*-[(ethylamino)methoxy]-

As used here, the terms 'part' and 'group' are not synonymous. Thus, '(dimethylamino)' is a two-part group even though it consists of three components; '(ethylamino)' is a two-part group consisting of two components. When a two-part group is attached to another group or radical, a three-part group is formed, and brackets are used to enclose the entire group name. Brackets are added for each successive part in a complex group name.

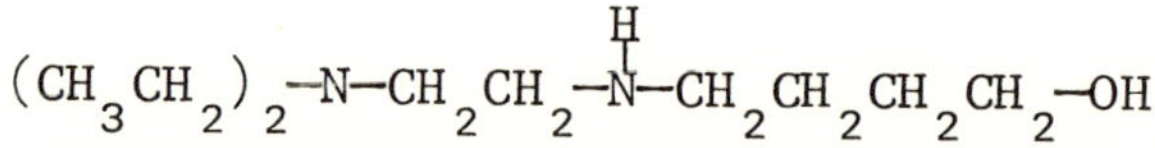

1-Butanol, 4-[[2-(diethylamino)ethyl]amino]-

As was stated earlier, an exception to the foregoing is the case in which a compound is named using the class names 'ketone' or 'ether.' In such a case, the spacing that results from the absence of final hyphens takes the place of the final set of curves or brackets.

Inorganic Compounds

Organic nomenclature has been discussed in some detail because it causes the most difficulty. By comparison, the naming of inorganic substances is relatively straightforward.

CA follows the customary practice of giving the more electropositive part of a simple inorganic compound first and the more electronegative part second. For compounds containing complex anions, however, additional entries are made for the sake of being as comprehensive as possible. For instance, the compound $Na_2[Ni(CS_3)_2]$ would be indexed both at Sodium bis(trithiocarbonato)nickelate(II) and at bis(trithiocarbonato) under the general heading Nickelates(II).

Those readers who are interested in a more general account of inorganic nomenclature than that provided by *The Naming and Indexing of Chemical Compounds* are referred to the sixty-seven-item bibliography, covering twenty-one languages, which has been published by Trimble.[13]

SOME SPECIAL CONSIDERATIONS

Several topics of varying importance have not been discussed in preceding sections, either because they did not fit naturally with the material being discussed or because they were considered beyond the scope of this book. What follows is a series of brief presentations which either elucidate such a topic or refer the reader to additional reading.

Special Prefixes

Confusion is sometimes caused by the use of differing prefixes which denote the replication of groups in a structure. The basic rule observed by *CA,* however, is simpler than many people realize. For simple groups the usual prefixes di-, tri-, penta-, etc., are used. For more complex groups, namely those which require curves, the prefixes bis-, tris-, tetrakis-, pentakis-, etc., are used.

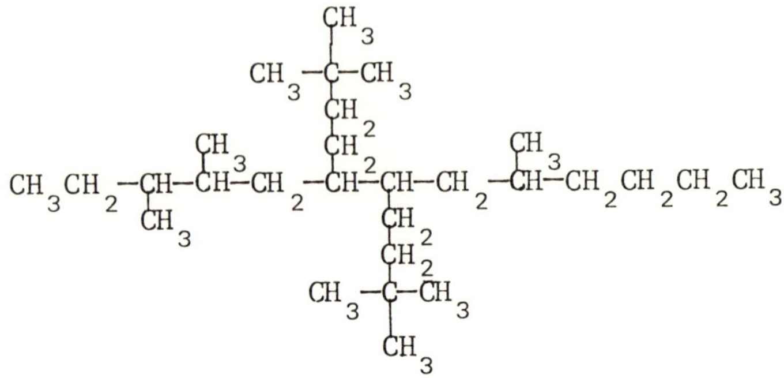

Tridecane, 6,7-bis(3,3-dimethylbutyl)-3,4,9-trimethyl-

The numerical prefixes determine alphabetical order only when they occur as the first letters within curves. Thus, bis(dimethylbutyl) precedes trimethyl because *d* comes before *m*.

The prefixes *sec-* and *tert-* stand for secondary and tertiary, and they take no part in determining the alphabetical order of substituents unless both are represented in the same compound. When this occurs, *sec-* precedes *tert-* but the group name to which they are affixed retains its alphabetical location with respect to all other substituent names.

_____, 6-*sec*-butyl-4-*tert*-butyl-3-methyl-6-vinyl-

Markush Structures

It is often useful to be able to represent structures which are partially known or to generalize by using "generic" substituent names for a series of compounds. This is usually accomplished by the use of Markush structures, which typically employ the symbol 'R-' to stand either for an unknown substituent or for one whose identity is deliberately being obscured. The former case is common in natural products work when the structure of a compound has been only partially elucidated. The latter case is typical of the patent literature, which often has a vested interest either in generalizing a concept or in obscuring the true nature of a process because of proprietary interests.

Names for Markush structures are usually constructed by using generic radical or group names such as 'alkyl' and 'aryl.'

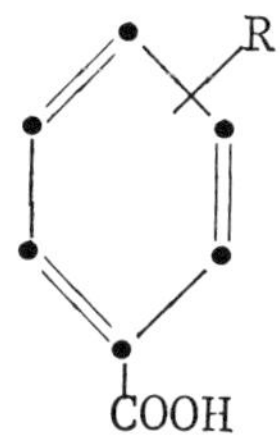

Benzoic acid, alkyl-

Stereochemistry

Basic stereochemistry, such as isomerism about a double bond and optical isomerism, is adequately covered by most chemistry textbooks, and advanced topics are beyond the scope of this book. Readers needing a good, nontrivial introduction to this topic are referred to the work of Eliel.[14] Those interested in certain specific recent developments might wish to read the work of Blackwood, et al.,[15] or that of Rush and White.[16]

Special Classes of Compounds

For certain classes of compounds such as alkaloids, carbohydrates, steroids, terpenes, and others, there is no substitute for the reports of IUPAC or the various American Chemical Society committees. A bibliography of these reports, which are available from the Chemical Abstracts Service, is contained in the *Handbook for Authors.*[17] These reports are either free or inexpensive.

Polymers are interesting in that they occur among many different types of compounds. The usual approaches to indexing them are by their structural repeating units or on the basis of their monomers. *CA*'s approach to this problem, which is based on the naming of monomers, has been discussed by Loening, et al.[18]

CONCLUSION

Considering its importance, chemical nomenclature has probably not received the attention it deserves. Although they are not so compact, systematic names are preferable to other kinds of linear codes in several respects. They, too, can be made unique and unambiguous; they do not need to be interpreted by or for practicing chemists; they are sufficiently compact for direct use as index entries; and they are suitable for oral discourse, which is clearly not the case with the other linear codes.

Chemical nomenclature also offers far better prospects for automatic processing than natural language text, because one knows in advance that only nouns will be encountered. This precludes all difficulties arising from ambiguous subject-predicate relationships. It was demonstrated as early as 1962 that systematic nomenclature is amenable to manipulation by the modern tools of language analysis: in that year Garfield published his work on nomenclature translation, a process involving the analysis of chemical names for machine conversion to appropriate empirical formulas.[19] A few years later Conrow demonstrated the feasibility of using the computer to generate the Baeyer names for certain bridged ring systems.[20] More recently, there have been publications dealing with the possibility of automated vocabulary control,[21] the conversion of systematic names into atom-bond connection tables,[22] and the linguistic analysis of chemical structure diagrams.[23]

There has been much emphasis on research and development using linear notation because of the advantages which accrue to brevity. However, as computer technology continues to improve, brevity may cease to be so important. In addition, with the advent of higher-level programming languages such as PL/I, which greatly facilitate character manipulation, it is probable that more researchers will investigate the use of the computer for handling chemical names.

Those who wish to know more about chemical nomenclature, but who are put off by the definitive reports, are referred to the eminently readable introduction by Cahn.[24]

NOTES

1. *Handbook for Chemical Society Authors,* 2d ed. (London: The Chemical Society, 1961), p. 48.

2. For a fascinating account of this work, see T. M. Lowry, *Optical Rotatory Power* (New York: Dover, 1964).

3. H. M. Leicester and H. S. Klickstein, *A Source Book in Chemistry* (New York: McGraw-Hill, 1952).

4. Fred A. Tate, "Progress Toward a Computer-Based Chemical Information System," *Chemical & Engineering News* 45 (1967): 78-90.

5. K. L. Loening, "International Cooperation on Scientific Nomenclature," *Journal of Chemical Documentation* 10 (1970): 231-236.

6. *Handbook for Authors* (Washington, D.C.: American Chemical Society, 1967).

7. See note 1.

8. *Chemical & Engineering News* 49, no. 42 (1971): 22.

9. Chemical Abstracts Service, *The Naming and Indexing of Chemical Compounds* (Easton, Pa.: American Chemical Society, 1962). [The Chemical Abstracts Service has recently made available new rules for the selection of names for the Ninth Collective Index Period. The application of these rules will result in index names which are more systematic than those appearing in indexes prior to this time.]

10. Ibid., p. 33n.

11. Charles H. Davis, "A Simple Code for Improving the Retrieval of Information Associated with Keto-Enol Tautomers," *Journal of Chemical Documentation* 6 (1966): 199-205.

12. Chemical Abstracts Service, *The Naming and Indexing,* p. 12n.

13. R. F. Trimble, "Bibliography of Rules of Chemical Nomenclature in Various Languages. I. Inorganic Chemistry," *Journal of Chemical Documentation* 10 (1970): 227-230.

14. Ernest L. Eliel, *Elements of Stereochemistry* (New York: Wiley, 1969).

15. J. E. Blackwood, C. L. Gladys, A. E. Petrarca, W. H. Powell, and J. E. Rush, "Unique and Unambiguous Specification of Stereoisomerism about a Double Bond in Nomenclature and Other Notation Systems," *Journal of Chemical Documentation* 8 (1968): 30-32.

16. J. E. Rush, and L. J. White, "Resolution of Ambiguities in the Nomenclature of Spiro Ring Systems," *Journal of Chemical Documentation* 10 (1970): 195-204.

17. *Handbook for Authors,* pp. 50-54.

18. K. L. Loening, W. Metanomski, and W. H. Powell, "Indexing of Polymers in *Chemical Abstracts,*" *Journal of Chemical Documentation* 9 (1969): 248-251.

19. Eugene Garfield, "An Algorithm for Translating Chemical Names to Molecular Formulas," *Journal of Chemical Documentation* 2 (1962): 177-179.

20. Kenneth Conrow, "Computer Generation of Baeyer System Names of Saturated Bridged Bicyclic, Tricyclic, and Tetracyclic Hydrocarbons," *Journal of Chemical Documentation* 6 (1966): 206-212.

21. Charles H. Davis, "An Approach to Automated Vocabulary Control in Indexes of Organic Compounds," *Journal of Chemical Documentation* 7 (1967): 131-134; and "An Approach to Automated Vocabulary Control in Indexes of Organic Compounds, II," *Journal of Chemical Documentation* 9 (1969): 252-256.

22. G. G. Vander Stouw, I. Naznitsky, and James E. Rush, "Procedures for Converting Names of Organic Compounds into Atom-Bond Connection Tables," *Journal of Chemical Documentation* 7 (1967): 165-169.

23. Kirk Rankin, and Stephen J. Tauber, "Linguistics as a Basis for Analyzing Chemical Structure Diagrams," *Journal of Chemical Documentation* 11 (1971): 139-141.

24. R. S. Cahn, *An Introduction to Chemical Nomenclature,* 2d ed. (Washington, D.C.: Butterworths, 1964).

8

Chemical Structure Representation: Linear Notation

In the preceding chapter we discussed chemical nomenclature, a method of describing chemical structures. This method can be viewed as an example of a *linear notation* system because a chemical name is probably always written in linear fashion (i.e., as a string or one-dimensional entity), no matter how complex the structure a chemical name purports to describe.

In this chapter, those representations of chemical structure that are popularly called linear notations are described. Some general information concerning linear notations is presented first, followed by a brief discussion of the origin and development of five well-known linear notation systems: IUPAC; Wiswesser; Hayward; Skolnik; and GREMAS. The way in which compounds are denoted within each system is also discussed and illustrated.

DEFINITIONS

A brief discussion of the typesetter's art will help elucidate the term linear notation. (What we shall say is applicable equally to such techniques as typesetting, typewriting, and printing through the medium of a computer line-printer.) When setting type, whether by the monotype or linotype process, the notion of a line is defined by the height of the body of each type element. Thus, if 8-point type is used, the line would be defined as one of arbitrary length whose height was 8 points (8/72 inch).[1] All symbols to be printed without special processing or handling would therefore have to fit within this "line." This notion of line is illustrated in Figure 32.

With this discussion of "line" as background, a *linear notation* can now be defined as any representation of chemical structure which is written as a string composed entirely of symbols that, in height, occupy one line of space. Table 2 presents several examples of linear notations.

Linear notations represent the elements of a language, just as words represent the elements of, for example, English or German. But to be effective as a means of communication, linear notations must be de-

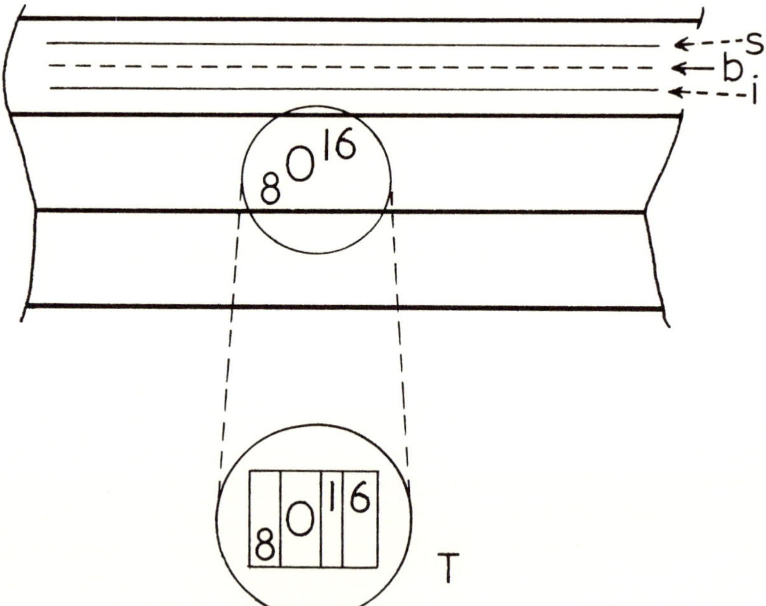

Figure 32–Illustration of the Notion of "Line"

s = center of superscript-character line; b = center of base line; i = center of inferior-character line; T = illustration of the placement of characters on the type slug.

Table 2–Various Linear Notations Representing the Compound "Toluene"

Notation	Structural Formula
Toluene	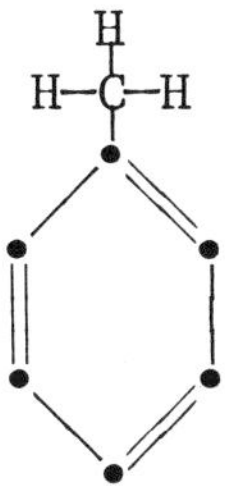
C_7H_8	
$C_6H_5CH_3$	
1B	
B6C	

rived and interpreted in a systematic way. Hence, a set of notations, together with the rules for their derivation and interpretation, is termed a *linear notation system.* Many such systems have been devised and have been used for a variety of purposes. Several of the currently popular systems will be described in this chapter. Before doing so, however, a few comments on the purposes and general characteristics of linear notation systems will be given.

PURPOSES

Not all of the linear notations given in Table 2 are of equivalent descriptive power. Toluene, for example, is a trivial name from which one cannot extract the structural formula by exercise of logical principles. As a second example, the molecular formula C_7H_8 represents more than one structural formula. But, according to the rules (logical principles) for forming and interpreting them, the remaining three notations are in one-to-one correspondence with the structural formula, and vice versa. That is, within their respective systems, these three notations uniquely represent the structural formula attributed to the substance we call toluene. The *uniqueness* of a representation is not, however, the only matter of concern to us. There is also the matter of *ambiguity.* Since C_7H_8 may be interpreted as representing more than one compound, it is said to be an ambiguous representation of "toluene." All the other examples of linear notations given in Table 2 are unambiguous within their respective systems because there exists only one interpretation of each notation as a representation of chemical structure.

Thus, two important characteristics of linear notation systems can be identified and defined:

> Uniqueness–The necessary and sufficient condition for *uniqueness* of a notation is that one and only one notation be derived from a chemical structure upon application of the rules of the system.
>
> Ambiguity–The necessary and sufficient condition for a notation to be *unambiguous* is that the notation be susceptible of one and only one interpretation upon application of the rules of the system.

These characteristics allow for the division of linear notation systems into four classes, as illustrated in Table 3. The choice of a notation system from among these classes of systems depends upon the purpose to which the notations are to be put.

Conservation of storage space is a major reason for employing linear notations as representations of chemical structure. But linear notations also have the important function of facilitating the storage and retrieval of chemical structural data so that the literature concerning such data is made more readily accessible to the chemist. In other words, linear notation systems serve as communication systems that link chemists who might otherwise be unaware of one another's existence. We may also be concerned with the precision and flexibility which such com-

Table 3–Classes of Linear Notation Systems

Class	Unique	Ambiguous
1	YES	NO
2	YES	YES
3	NO	NO
4	NO	YES

munication systems allow. For instance, a linear notation system that yields unique and unambiguous notations allows for precise communication about specific substances, but it may render communication about classes of substances more difficult. In contrast, a linear notation system that yields nonunique, unambiguous notations has the merit of easier application (by virtue of reduced rigor in the rules for deriving notations), but it usually requires more work to communicate information on classes of structures. Moreover, its lack of uniqueness means that stores of such notations will contain redundant entries. Nevertheless, in particular situations it is conceivable that any of the four classes of linear notation system might be justifiable. The choice depends upon a variety of factors, including cost, number of structures to be described, number of chemists to be served by the system, growth of the number of structures, expected number of specific *versus* class queries to be serviced, whether manual or computer-based retrieval is to be provided, desired cost/benefit ratio, and so on.

As we have said, linear notation systems may yield notations that facilitate retrieval of data on either specific substances or classes of substances. The developer of a linear notation system frequently attempts to accomplish both goals by means of a single system. Let us consider some of the factors involved in developing a "general purpose" notation system.

A linear notation system that provides for the derivation of notations which correspond symbol-by-symbol to the atoms and bonds in a structural formula will certainly satisfy the requirement of specificity. For example, valeric (pentanoic) acid might be represented by a notation of the following type:

Structural Formula	**Notation***
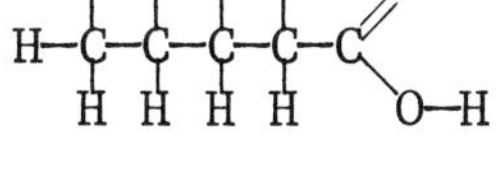	CCCCC(=O)O

*Hydrogen is assumed to fill "empty" valences.

But what facility does such a notation afford for class retrieval? Suppose we wished to retrieve data on carboxylic acids. To be sure, the identification of the symbol pattern C(=0)0 would serve for the retrieval of all carboxylic acids, but for this purpose a more efficient* notation may be devised, one in which the carboxylic acid group is represented by a single symbol, say X. For example,

CCCCX

In fact, for purposes of retrieval by classes of structure, many linear notation systems have been developed which employ this device. Most of these systems yield notations that are both nonunique and ambiguous. One example is the principal linear notation system used at the Smith, Kline & French Laboratories (SK&F).[2] In this system a chemical structure is represented by means of symbols which usually correspond to reactive (functional) groups (*cf.* Chapter 7) in each particular structure, much as the chemist has dealt with such groups over the years. For instance, the name "sulfanilamide" has its origins in a desire to describe the active groups in the compound

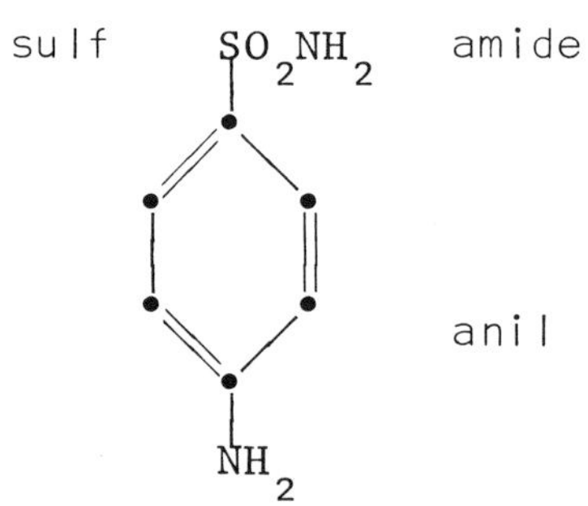

which, in more systematic terms, might be called 4-aminobenzenesulfonamide. In the SK&F system, three "groups" would be singled out for notational representation: NH_2, 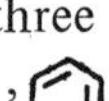 and SO_2NH_2. Each group is

*Efficiency here is based on the number of symbols which must be examined in order to identify the desired entity.

assigned an alphanumeric descriptor, e.g., F5L1 for NH_2, so that, on retrieval, one need only specify F5L1 to retrieve compounds involving NH_2 attached to a benzene nucleus. Of course, the absence of positional indicators would allow the three isomers orthanilamide, metanilamide, and sulfanilamide to be represented by a single notation, e.g., 1771–F5L1–NYR1, corresponding to SO_2NH_2 NH_2 .

Such notations can be manipulated manually, by means of relatively simple machinery such as sorters and collators, or by means of computers. They have the advantage of simplicity, and provide for class retrieval. When the number of structures represented by such notations is on the order of a few thousands, the notations serve adequately, though imprecisely, for retrieval of specific structures.

Sulfanilamide illustrates a principle that permits us the transition from nonunique, ambiguous linear notations to those in which the developer has striven for greater systematization and perhaps uniqueness of representation as well. We observe that sulfanilamide has two $-NH_2$ groups in its constitution, and although, as chemists, we are aware of the (usually) different reactivities of the NH_2 groups in different environments, it may be desirable to represent the NH_2 group by the same notational symbol, regardless of the relation it bears to other groups in a particular structure. In the Wiswesser Linear Notation System, for example, provision is made for representing the NH_2 group by the symbol Z, regardless of the environment in which the group is found. Similarly, the sulfonyl group, SO_2, is represented by SW regardless of its surroundings.

The main advantage of such systematization is that notations may be generated using a small set of symbols (some 40 or so symbols are used in the Wiswesser Linear Notation System), whereas a notation system of the type employed by Smith, Kline & French requires a large set of symbols (some 700 or so actually being used by SK&F). On the other hand, the benefit derived from more regularized notation systems is at least partially offset by the increased complexity of the rules required for the derivation and interpretation of the notations. This point will be elucidated later in this chapter, but it is noted here that the regularization of notation is epitomized by the system of structural formulas themselves. Through the use of 105 element symbols, together with some half-dozen "linkage" symbols (-, =, ≡, ←, +, −, etc.), the chemist is able to represent, uniquely, every chemical species whose structure is known to him. Thus, one might have as his basic goals the development of a linear notation system whose descriptive capabilities were equivalent to those of structural formulas but which required the use of much fewer symbols. We shall see how well these goals have been met in various instances, although we may observe here that the complexity of the rules for deriving and interpreting linear notations is in an inverse relationship with the number of symbols employed in the notation system. Thus, a compromise must usually be established between notational compactness and ease of use of the notations.

The developer of a linear notation system faces additional problems. As already noted, he must effect a compromise between compactness of notation and complexity of rules, assuming there is a high degree of regularization in the system. Another factor which must be considered in the development of a linear notation system is the use to which the notation will be put. Here "use" is employed in the sense of who will use the notations rather than for what purpose it will serve. For example, a linear notation system might be used for retrieval of classes of structures, but in such a way that manipulation of the notations is entirely computer based, so that the chemist never "sees" the notation and therefore does not need to know anything about its formation or interpretation. In such a case, the notation system could be designed to take advantage of the power of the computer to rapidly and accurately process very compact notations.[3]

On the other hand, if the manipulation of the notations is to be at least partly in the hands of humans, then the developer of the notation system would do well to realize that the human is the weak link in his system and that the rules for deriving and interpreting the notations should accommodate the human at the expense of notational compactness.

Another factor which must be accounted for in the design of a linear notation system is that involving the conflict between generic (structural-class) retrieval and specific (individual-structure) retrieval. Several parameters must be carefully balanced if one is to satisfy basic design criteria. The choice of symbols which correspond one-to-one with structural formulas gives maximum specificity, no generic capability, and great compactness, as exemplified by the assignment of an arbitrary number to each structure. But when a very small number of symbols is used, which must be combined to represent a given structural formula, the generic capability increases dramatically, specificity falls to near zero, and compactness increases somewhat. Similarly, many other combinations of generality, specificity, and compactness can be achieved. With these relationships in mind, it should be easier to make choices commensurate with system design criteria.

Still another factor to be considered is that related to the types of structures to be represented. A linear notation system designed to represent only organic compounds should be simpler than one designed to handle all types of chemical structures. Similarly, a notation system designed to represent only linear structures (tree structures in graph-theoretical terminology) should be easier to devise than one designed to represent both linear and nonlinear (cyclic) structures. It will be seen subsequently that some existing notation systems facilitate representation of one type of structure or another, even though inadvertently so.

Much of the foregoing discussion of design criteria is summarized in a list, published in 1949, of desiderata for an internationally acceptable linear notation system.[4] These desiderata include:

(1) Simplicity of use

(2) Ease of printing and typesetting

(3) Conciseness (compactness)

(4) Recognizability (ease of interpretation)

(5) Capability of producing a unique chemical nomenclature

(6) Compatibility with inorganic nomenclature

(7) Uniqueness (same as 5)

(8) Capability of producing unambiguous and useful notations

(9) Ease of machine manipulation

(10) Descriptiveness (i.e., capable of exhibiting structural association)

(11) Capability of handling descriptions of incompletely defined structures

In essence, what was required was a linear notation system capable of dealing with all classes of compounds and producing unique and unambiguous notations that were easy to generate and to interpret, to which were adduced the criteria of descriptiveness, conciseness, and ease of machine manipulation. Such criteria have recently been reiterated by the Committee on Chemical Information.[5] The committee stressed the importance of dealing with substances whose structures were incompletely delineated (stereochemistry being one such structural feature giving rise to incomplete specification).

So far we have seen how linear notation systems came to be developed, why they were developed, and the factors which influence their design. We turn now to a description of several specific linear notation systems.

SPECIFIC NOTATION SYSTEMS

A study of the literature shows that linear notation systems are frequently discussed, compared, and evaluated with respect to their *indexing* capabilities. In the discussion that follows, we are concerned only with how the various systems meet the criteria for structural representation set forth by their designers. Indexing considerations will be treated separately, but it is observed here that a printed index is likely to serve the needs of all its users in only a very few instances and that, in a computer-based retrieval system (index), by contrast, the only requirement for servicing the needs of all its users is an adequate representation of structure. Thus, the most important factor to consider in designing a linear notation system is its capacity for representing, within system-design constraints, the chemical species with which one wishes to deal. Since these criteria have already been discussed, we will proceed directly to a discussion of specific linear notation systems. The systems to be considered are:

(1) The IUPAC or Dyson System

(2) The Wiswesser System

(3) The Hayward System

(4) The Skolnik System

(5) The GREMAS System

The IUPAC Notation for Organic Compounds

In 1946 Dyson announced the development of a notation system for organic compounds in a lecture before the Royal Institute of Chemistry (London).[6] As originally conceived, the notation system provided for the derivation of notations consisting of symbols selected from the upper case Roman alphabet, the ten Arabic numerals, and ten punctuation marks. (In symbolism, this system bore a remarkable similarity to that employed later in the Wiswesser notation system.) As will be seen, the rules for producing notations paralleled to a considerable degree the rules for systematic chemical nomenclature. The Dyson system represents the earliest attempt to develop a systematic notation for organic compounds as a successor of, or adjunct to, systematic chemical nomenclature.

Shortly after Dyson published his work on a notation system, the IUPAC appointed a Commission on Codification, Ciphering and Punched Card Techniques to study and report on the various notation systems then available for organic structures.[7] In 1949 the commission set forth the list of desiderata that was presented, in modified form, earlier in this chapter. The adoption of a provisional international linear notation system based upon the Dyson system was recommended in a report published in 1951.[8] Seven years later a tentative draft of the international system was printed for review by the organizations adhering to the IUPAC and, finally, in 1961, the definitive *Rules for IUPAC Notation for Organic Compounds* was published.[9] The IUPAC notation system has undergone no change since 1961, although Dyson has suggested several modifications which would improve its use by simplifying the notations or by extending the range of substances that can be represented by such notations.

As mentioned earlier, the basic design of the Dyson (hence, the

IUPAC) notation system followed rather closely the principles of systematic organic chemical nomenclature. Thus, in notations for acyclic and cyclic structures precedence is given to the carbon skeleton rather than to reactive, or functional, substituents. For example:

Structure	IUPAC Notation
1. $CH_3CH_2CH_2CH_2CH_2COOH$	C_6X
2. $HOOCCH_2CH_2CH_2COOH$	C_5X15

Similarly, functional groups are cited in an IUPAC notation according to rules of precedence somewhat like those used in chemical nomenclature. Priority is given to the symbols X (for carboxylic acid), EQa (aldehyde), EQ (ketone), and Q (alcohol), in that order. All other groups are cited in descending lexicographic order.* The following examples illustrate both the use of the symbols and the ordering principle:

3. CH_3COOH	C_2X
4. CH_3CHO	C_2EQa
5. $CH_2{=}C{=}O$	C_2E1EQa
6. CH_3CH_2OH	C_2Q
7. $HOCH_2CH_2CHO$	$C_3EQa1Q3$
8. $OHCCH_2CH_2COOH$	$C_4X1EQa4$

*A set of symbols in lexicographic order is a well-ordered set. A well-ordered set is defined as a set such that it and its every nonempty subset has a first element, and such that for any two elements a and b of the set, either a precedes b or b precedes a. If the ordering begins with the first element (that element which has no predecessor), then we call this order *ascending*. The reverse order is called *descending*.

Let us now state the general principles upon which all notations are based in the IUPAC system. No attempt is made to give a comprehensive statement of principles, but rather the intent is to show, in general, how organic compounds are viewed in the IUPAC system.

As has already been stated, preference is given the carbon skeleton in forming IUPAC notations, except that ring systems, whether carboxcyclic or heterocyclic, are given precedence over acyclic systems when these occur together in a structure. Otherwise the cyclic and acyclic components are treated independently.

Acyclic Structures

In acyclic systems, the longest chain of atoms is selected to form the initial part of the notation, and branches and functional substituents are cited subordinate to the main or longest chain. The atoms of the main chain are numbered so that branches and functional substituents are attached to atoms of the main chain with the lowest possible numbers. This is analogous to the low-numbering principle used in chemical nomenclature, which is described in Chapter 7. Branches and substituents are numbered, where necessary, in an analogous fashion. In the notation, branches are cited in descending order of priority, i.e., in order of decreasing size. The following examples illustrate how notations are derived for acyclic hydrocarbons.

9. CH_4 — C

10. $CH_3CH_2CH_3$ — C_3

11. $CH_3(CH_2)_{18}CH_3$ — C_{20}

12. $CH_3CH_2{-}C(CH_3)(CH_2CH_3){-}CH_2CH_3$ — $C_5C_2 3C3$

13. $CH_3CH_2CH_2CH(CH_3CHCH_3){-}CH_2CH_2CH_3$ — $C_7(2C_3)4$

14. 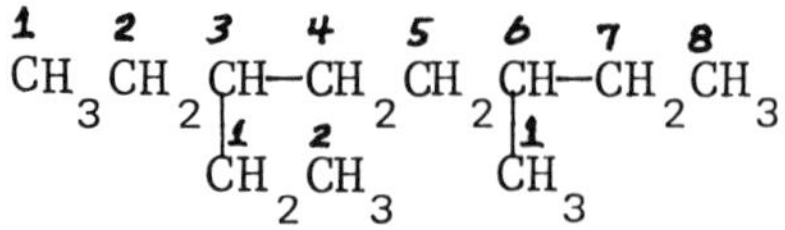

$C_8C_2 3C6$

15. 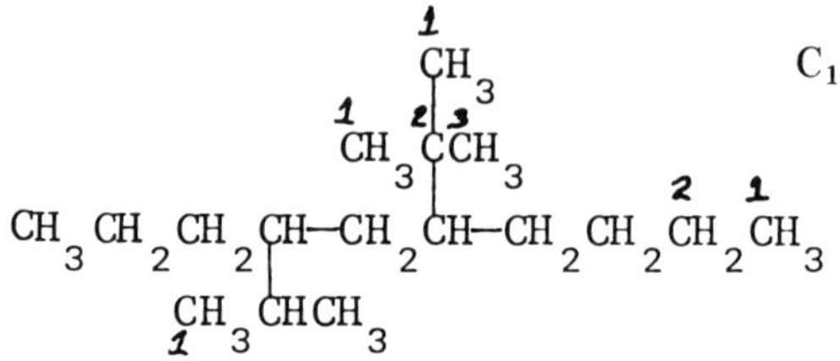

$C_{10}(2C_3C2)5(2C_3)7$

It will be seen that simple (unbranched) branches are numbered beginning at the atom attached to the main chain. Simple branches are cited in the notation without enclosing marks (parentheses) and the locant or point of attachment to the main chain is cited immediately following the branch symbol (see examples 12 and 14). For complex (branched) branches, the same rules are applicable, except that the branch is enclosed in parentheses and the locant of the branch atom attached to the main chain is placed first within the enclosing marks (see examples 13 and 15). Although complex branches could be represented by notations in which parenthesized expressions are nested as far as necessary to delineate the entire branch, they are instead represented in "assembly notation." Before discussing assembly notation, however, a few remarks concerning functional groups and unsaturation are in order. Considering unsaturation first, the IUPAC rules provide for representing a double bond by the symbol E and a triple bond by the symbol Y. (All unsaturation is so represented, except for that in aromatic systems. These systems are discussed later with cyclic structures.) Highest priority is given double bonds; then follow triple bonds and finally single bonds. The following examples illustrate these principles:

16. $CH_3CH_2CH_2CH_2CH_3$ C_5

17. $CH{\equiv}CCHCH_2CH_3$ C_5Y

18. $CH_2{=}CHCH_2CH_2CH_3$ — C_5E

19. $CH{\equiv}CCH_2C{\equiv}CH$ — C_5Y14

20. $CH_2{=}CHCH_2C{\equiv}CH$ — C_5E1Y4

21. $CH_2{=}CHCH_2CH{=}CH_2$ — C_5E14

22. 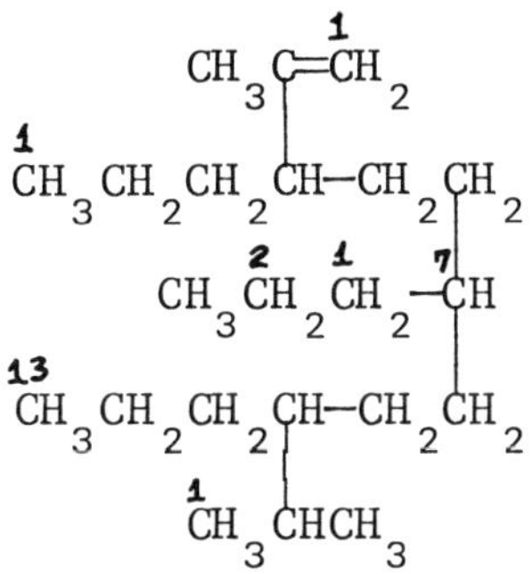

$C_{13}(2C_3E)4C_37(2C_3)\underline{10}$

(Arrow from 18 to 22: Increasing Priority)

The last example (22) illustrates both the choice of senior branch ($CH_3{-}\overset{|}{C}{=}CH_2$) and how the senior branch controls the numbering of the main chain.

In the IUPAC system, functional groups are treated in a manner somewhat analogous to that for branches. We have already mentioned some types of functional groups (examples 3–8). It will suffice for this discussion to state that substituents are cited much as they would be in molecular formulas or line formulas of the type we have been using under the heading "structure" in our examples. For single-atom substituents such as -Br, -Cl, -F, -I, and atoms with hydrogen, such as -SH, $-NH_2$, $-SiH_3$, the accepted atom symbol* is placed in the notation, together with a locant, according to the priority rules already alluded to. When the substituent is bound to a chain (or branch) by a double or triple bond, the symbols E and/or Y are used to denote this fact. The following examples illustrate how simple functional substituents are incorporated into IUPAC linear notations:

*Ch is used for chlorine to avoid confusion between the "one" (1) and the "ell" (l).

23. 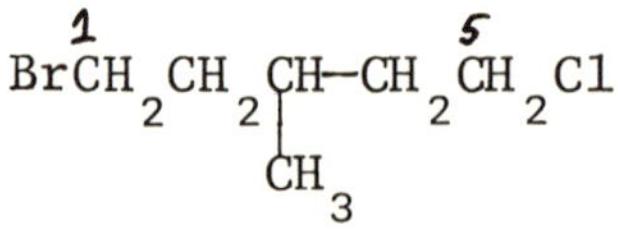 C_5C3Br1Ch5

24. 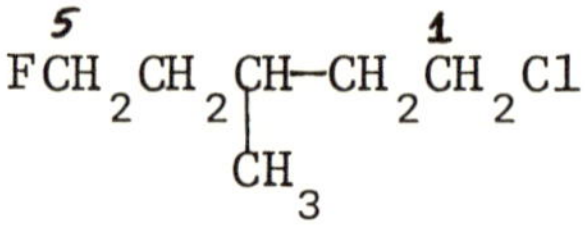 C_5C3Ch1F5

25. 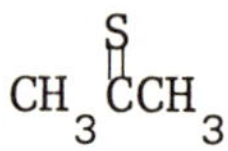 C_3ES2

26. $CH_3C{\equiv}N$ C_2YN

27. $BrCH_2CH_2CH{=}NH$ C_3Br1EN3

When the functional substituents are composed of two or more atoms (called, in the IUPAC system, *collective groups*), one of three procedures applies. For "oxygenated" functional groups, such as phosphate, sulfate, nitrate, nitro, nitroso, and iodoxy, the practice normally followed by chemists is used. Thus, IO, IO_2, NO, NO_2, PO, PO_2, SO, SO_2 are all correct symbol aggregates in IUPAC notation. For example,

28. $CH_3CH_2CH_2NO_2$ C_3NO_2

29. $CH_3(CH_2)_{10}CH_2SO_3H$ C_{12} SO_2Q

The second type of collective group consists of those functional groups of the form $(\mathrm{C})\langle^{=X}_{-Y}$, where the carbon is part of the main chain. This group is exemplified by the amides, acid chlorides, and amidines. Note that the semicolon (;) denotes that the symbol(s) following modify the same atom of the main chain as the preceding symbol(s). The point of attachment is denoted by a locant following the symbol(s) that come after the semicolon, if other than 1, and refer to the

30. $CH_3CH_2C(=O)NH_2$ — $C_3EQ;N$

31. $CH_3CH_2C(=S)-Br$ — $C_3Br;ES$

32. $(CH_3)_2CHCH_2CH_2C(=NH)NH_2$ — $C_5C2EN;N5$

complete aggregate of symbols, as in example 32 (where the terminal 5 means that both EN and N are attached to atom 5 of the main chain, thus EN;N5).

The third type of collective group is exemplified by compounds like urea, guanidine and semicarbazide, which contain an unsubstitutable carbon atom. The form of the notation representing these groups depends upon whether the parent compounds are being considered or the groups are substituted by a single carbon chain. The following examples illustrate the two forms of notation. If more than two chains are attached to a group in this category, assembly notation is used (see below).

33. $NH_2C(=O)NH_2$ — $CEQ;(N)_2$ *

34. $NH_2C(=O)NHNH_2$ — $CEQ;N;N_2$

35. $NH_2C(=NH)NH_2$ — $CEN;(N)_2$ **

36. $CH_3CH_2CH_2NHC(=S)NH_2$ — $C_3NCES;N$

*Contracted from CEQ;N;N
**Contracted from CEN;N;N

37. $CH_3CH_2CH_2NH\overset{O}{\overset{\|}{C}}NHNH_2$ $C_3NCEQ;N_2$

38. $CH_3CH_2NH\overset{NH}{\overset{\|}{C}}NH_2$ $C_2NCEN;N$

39. $CH_3CH_2N{=}C(NH_2)_2$ $C_2NEC;(N)_2$

Assembly notation is employed whenever a compound has a discontinuous carbon skeleton. Typical examples of such compounds are the ethers, esters, and nitrogen-substituted amides. The noncarbon atom or atoms which "interrupt" the carbon skeleton are called *links.* In general, a link is defined as any di- or polyvalent element other than carbon, or any unsubstitutable group of atoms. One special case, that of esters, will be considered separately.

When two carbon skeletons are joined by a link, the senior skeleton is cited first in the notation, followed by a colon, locant, link, and stroke (slash), in that order; then the other carbon skeleton is joined to the link. The following examples illustrate the application of assembly notation:

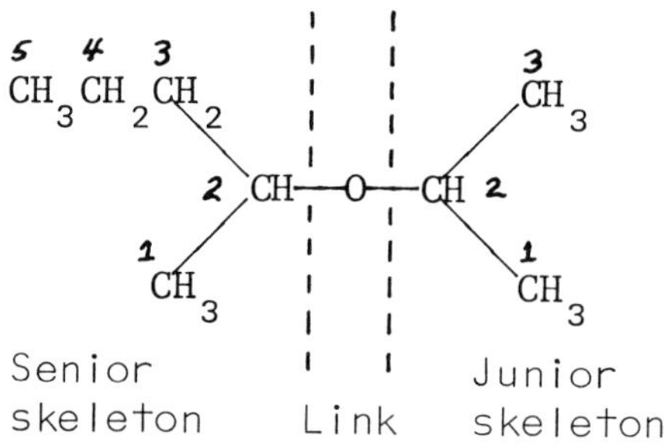

$C_5:2Q/2C_3$

40. $(CH_3)_2CH\underset{CH_2CH_3}{\underset{|}{CH}}(CH_2)_3S(CH_2)_2C(CH_3)_3$ $C_6C_23C2{:}6S/4C_4C2_2$

41. $CH_3(CH_2)_4N{=}N(CH_2)_2CH_3$ $C_5{:}NEN/C_3$*

42. $CH_3(CH_2)_3OCH_2CH_2SCH(CH_3)_2$ $C_4{:}Q/C_2{:}2S/2C_3$**

43. 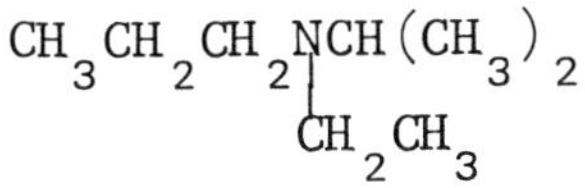 $C_3{:}N(C_2)/2C_3$

44. 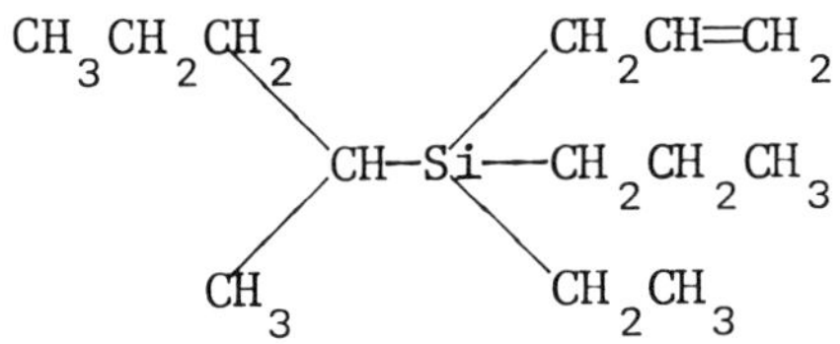 $C_5{:}2Si(C_3)(C_2)/3C_3E$

There are special subrules for deriving assembly notations, especially as regards symmetry about a link, but these will not be considered here. This discussion of assembly notation will be concluded with a brief consideration of esters.

Esters are accorded notations beginning with the senior component (whether derived from the alcohol or acid portion of the ester), followed by the symbol X and then the notation for the junior component. A stroke is placed *before* X if the carbonyl group is part of the junior component and *after* X if the carbonyl group is part of the senior component. Locants for X are placed on the side of X opposite the stroke. The following examples illustrate these rules:

45. 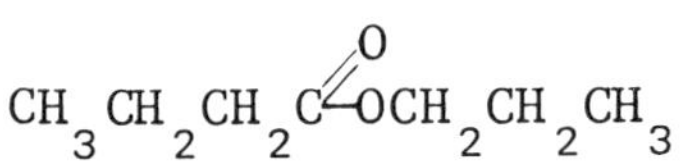 C_4X/C_3

46. 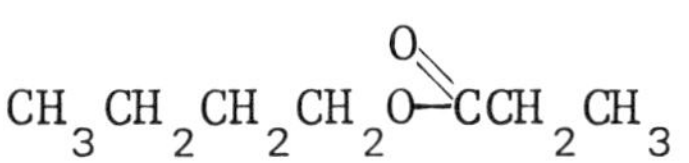 C_4/XC_3

*The locant "one" immediately before or after a colon or stroke is omitted.

**If the interposed carbon chain consists of just one unsubstituted carbon, the slash before and colon following this carbon are omitted.

47. 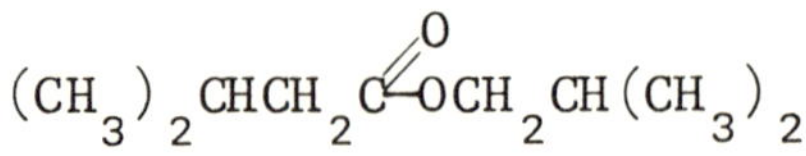 $C_4C2{:}4X/C_3C2$

48. 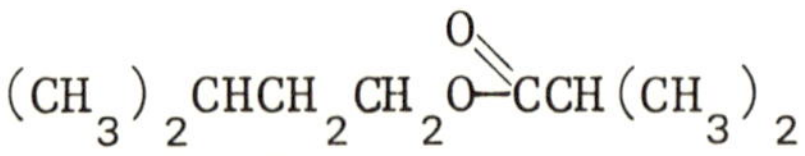 $C_4C2{:}4/X4C_3C2$

49. 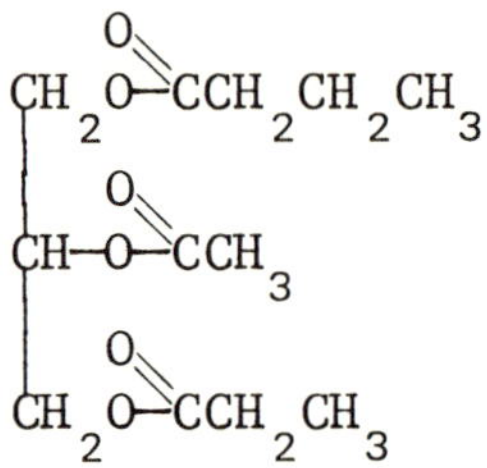 $C_4X/C_3(XC_2)2{:}3/XC_3$

Assembly notation is extended to structures with "uninterrupted" carbon skeletons when "secondary" parentheses would otherwise be required. This idea is illustrated below; further details regarding the application of assembly notation are best obtained from the definitive rules.[10]

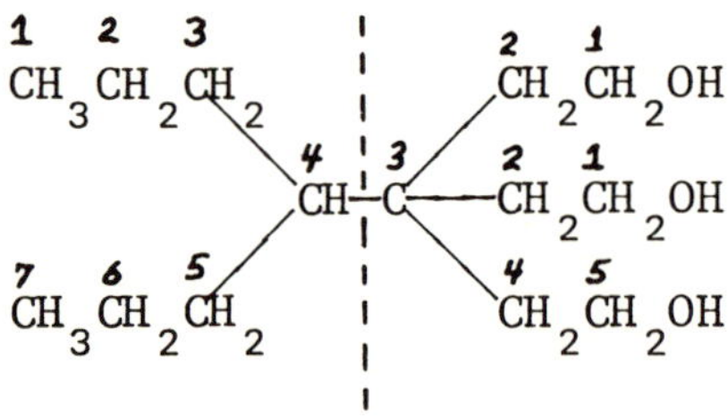

Preferred

$C_7{:}4/3C_5(2C_2Q)3Q15$

Not Preferred

$C_7(3C_5(2C_2Q)3Q15)4$

Cyclic Structures

Cyclic structures are assigned notations essentially independently of the notations derived for acyclic structures, except for derivation of locants and determination of seniority. Whenever both types of structure are present in a compound, cyclic components are given seniority over acyclic components. Simple (one-ring) carbocycles are assigned a basic notation consisting of a letter (either A or B), followed by a numeral indicating the number of atoms in the ring. For example:

50. 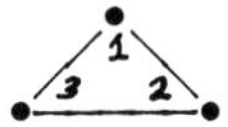 A3

51. 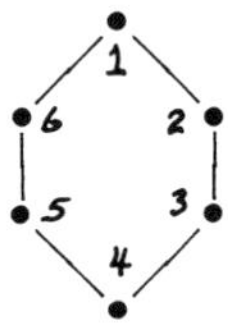 A6

52. 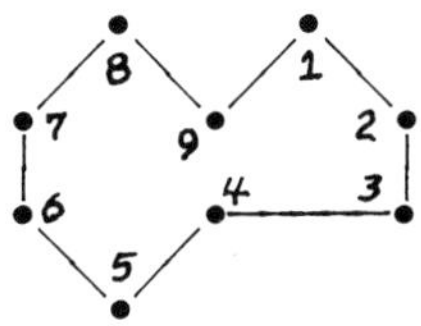 A9

53. 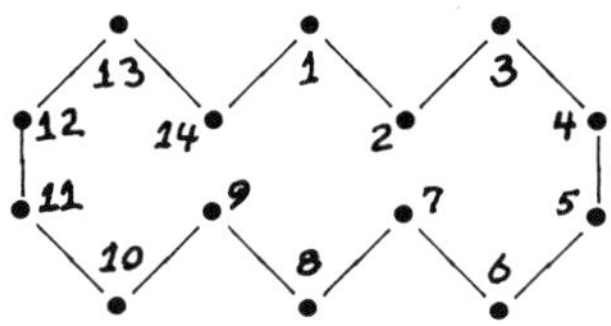 A14

Numbering of the ring atoms is sequential from any arbitrary starting point in either direction. When the ring is substituted, numbering is chosen so as to give lowest possible numbers to substituents and groups in descending order of seniority. For example:

54. 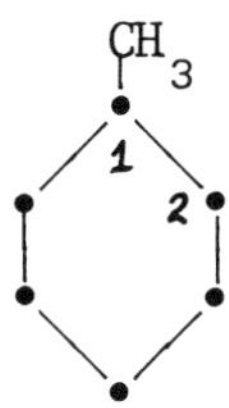 A6C

55. 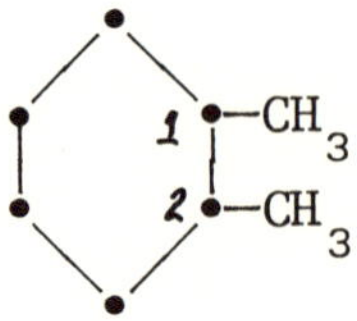

$A6C_2 12$

56. 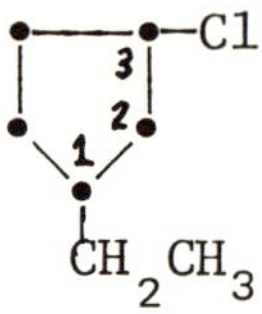

$A5C_2 1Ch3$

57. 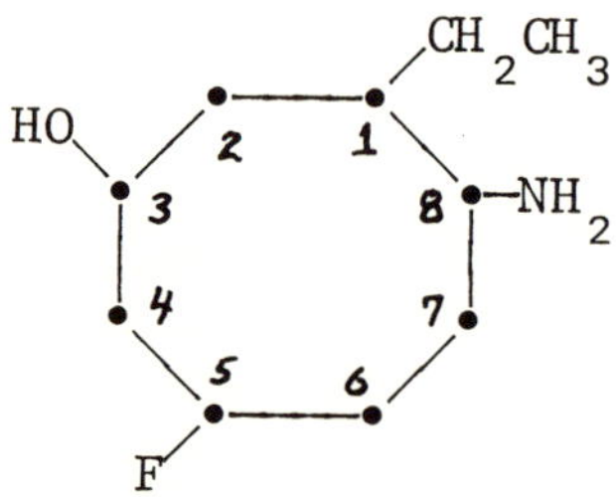

$A8C_2 1Q3F5N8$*

As for acyclic compounds, assembly notation is used whenever the carbon skeleton is interrupted (exclusive of noncarbon atoms in rings). For example:

58. 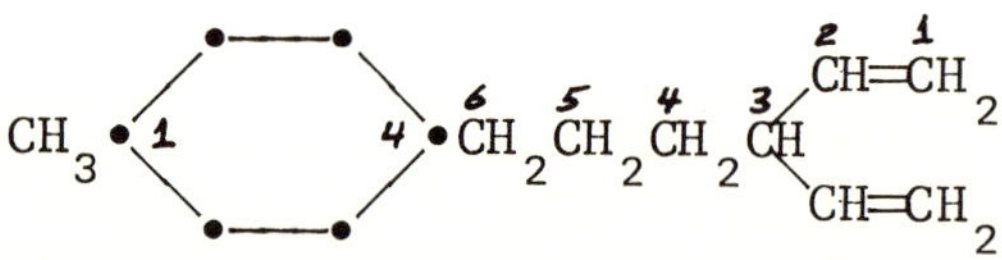

$A6C{:}4/6C_6(C_2E)3E$**

59. 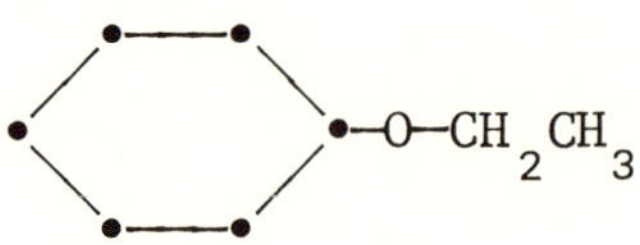

$A6{:}Q/C_2$

*Note that numbering of the ring atoms is prescribed by the seniority of the substituents. $C_2 > Q$, $Q > N$.

**Numbering of the ring is prescribed by assembly notation rules rather than branch seniority.

Unsaturation in cyclic structures is denoted by E, for double bonds, or Y, for triple bonds, except when the cycle contains the maximum number of noncumulative double bonds,* in which case the cycle is denoted by B followed by the number of atoms in the ring.** Consider the following examples:

60. 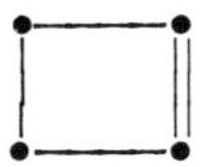 A4E

61. 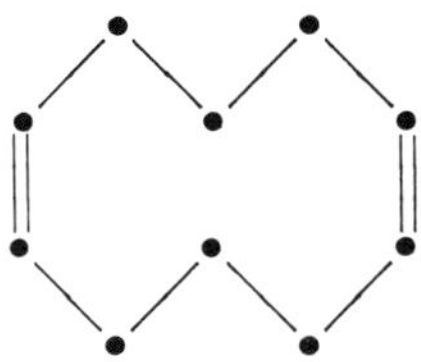 A1$\underline{0}$E16

62. 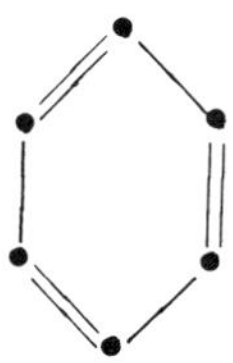 B6

63. 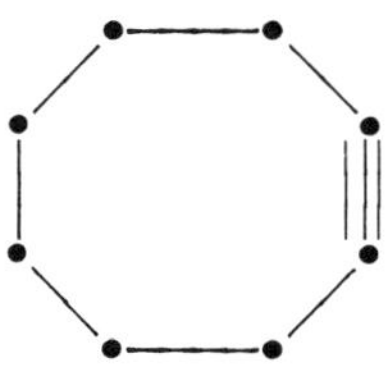 A8Y

*Systems of double bonds are categorized as cumulative, conjugated, or isolated, according as they join pairs of atoms separated by no, one, or more than one single bond, respectively.

**The "B" notation is used for those cycles which one usually thinks of as "aromatic." The notation is also used for cycles bearing substituents which render the extant number of double bonds fewer than the maximum number possible in the unsubstituted cycle.

64 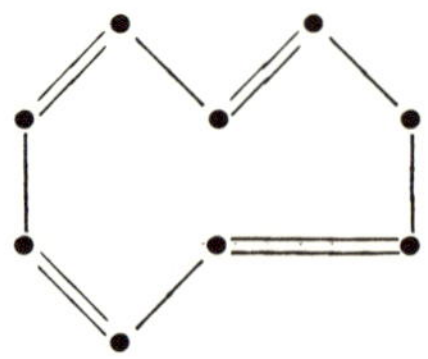 B9

When a ring, when unsubstituted, would be assigned the B notation but, because of substitution, contains fewer double bonds than its unsubstituted counterpart, *implied* (h), or *added* (H) *hydrogen* notation is used. The number of double bonds remaining in the ring must be greater than or equal to the number of ring atoms which do not share a double bond. Otherwise "A" notation is used. For example:

65. 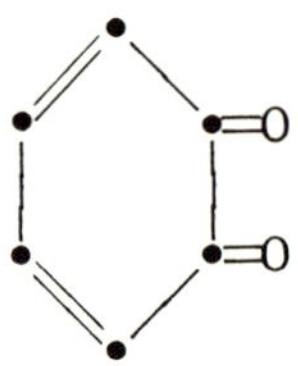 B6EQ12

66. 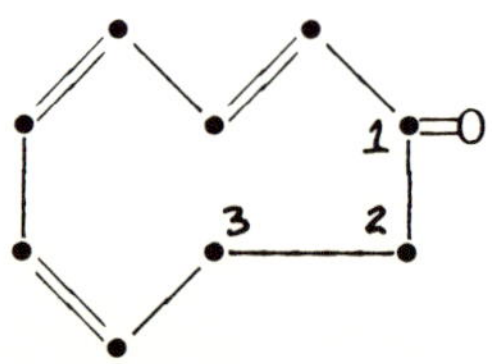 B9EQ1H23

67. 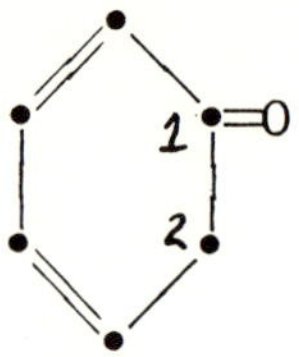 B6EQ1h2

68. 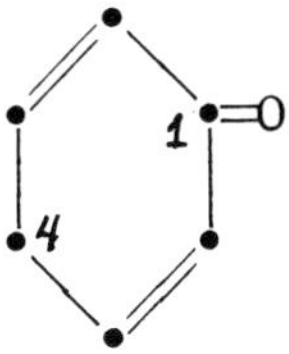

B6EQ1h4

69.* 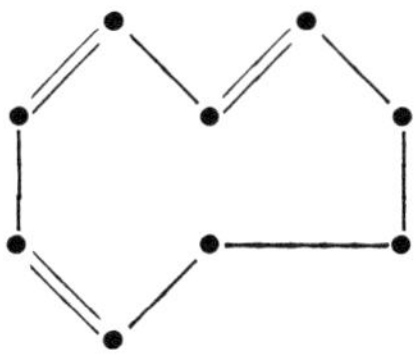

B9H23

70. 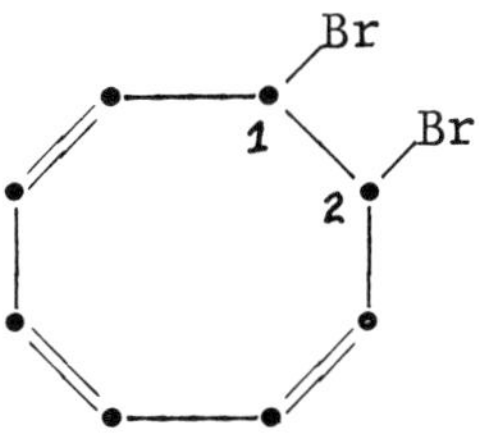

B8Br12H12

71. 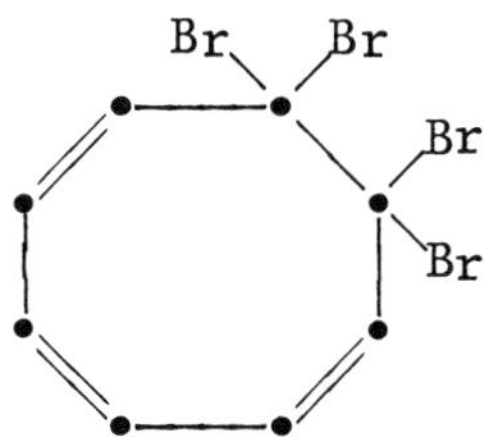

B8Br$1_2 2_2$

72.** 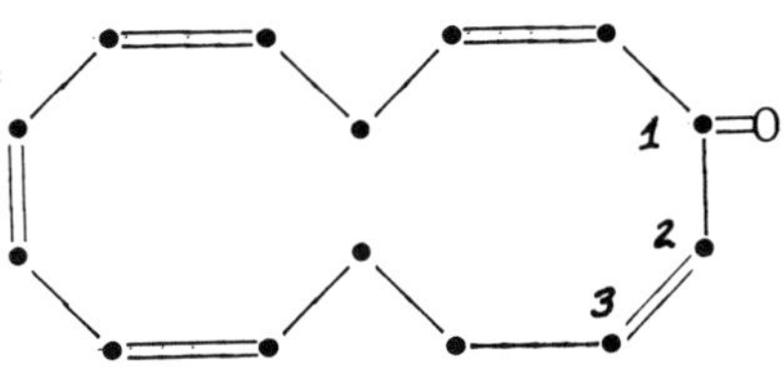

B1$\underline{4}$EQ1h4H51$\underline{2}$

*Here hydrogen is the substituting element (at positions 2 and 3).

**h is senior to H both in order of citation in the notation and in choice of low numbers for locants.

Heterocycles are handled analogously to carbocycles except that "Z" terms are added to the notation to denote the heteroatoms in the ring. The Z term immediately follows the ring term (i.e., the A or B term). For example:

73. 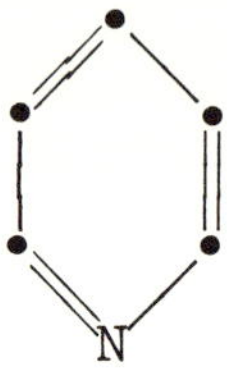

B6ZN

74. 

A5ZQ

75. 

B5ZS

76. 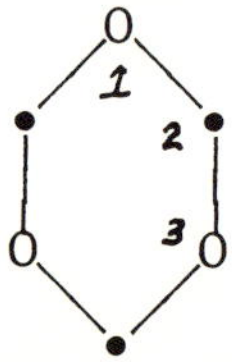

A6ZQ135

Z terms are senior to all substituents of whatever kind and to E and Y terms. When the ring has fixed numbering, locants must be used with single Z terms. Locants are also required for multiple Z terms. For example:

77. 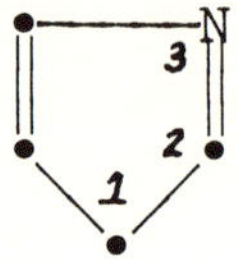

B5ZN3

78. 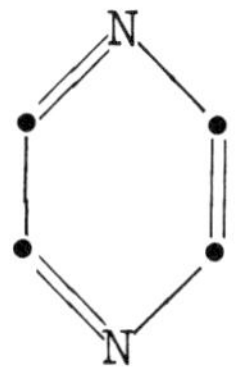

B6ZN14

79. 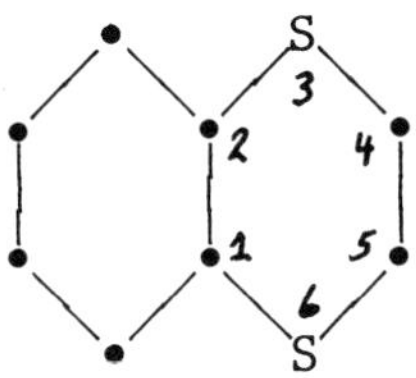

$A6_2ZS36$

This discussion of the IUPAC Notation System will be concluded with a consideration of complex ring systems. Ring systems are considered, for notational purposes, to be members of one of these categories: systems of isolated rings, systems of ring assemblies, or systems of ring aggregates. Examples of each of these types of system are presented in Table 4 (see page 170).

Each class may be defined as follows:

Type	Definition
Isolated	Rings linked by acyclic components
Aggregated	Two or more rings which have atoms and/or edges in common
Assembled	Two or more rings or aggregates of rings linked so that there are no intervening acyclic components

Both the aggregated and the assembled types are further divided into two classes: those having rings all of one size (called *homoreticular*) and those having rings of different sizes (called *heteroreticular*). In develop-

Table 4—Examples of Systems of Isolated Rings, Ring Assemblies, and Ring Aggregates

	Contains		
Structure	Isolated	Aggregated	Assembled
	Systems of Rings		
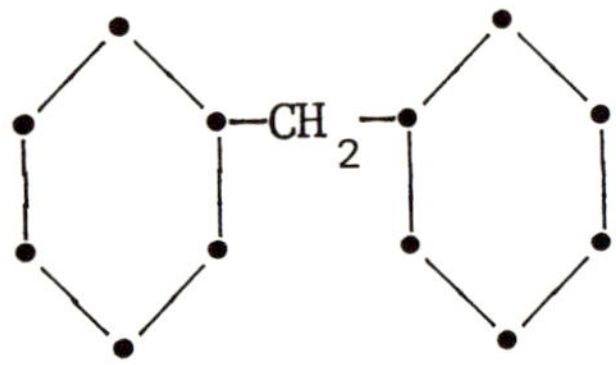	X		
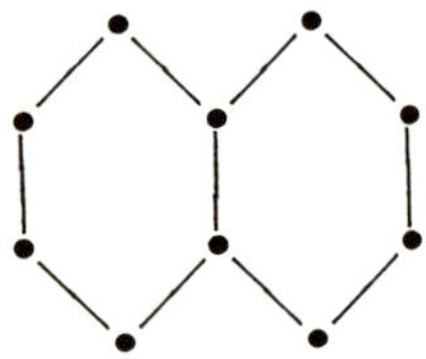		X	
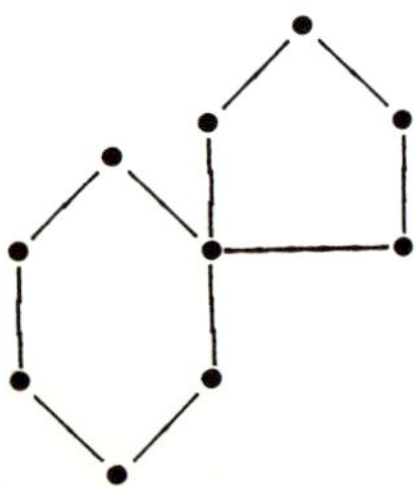		X	
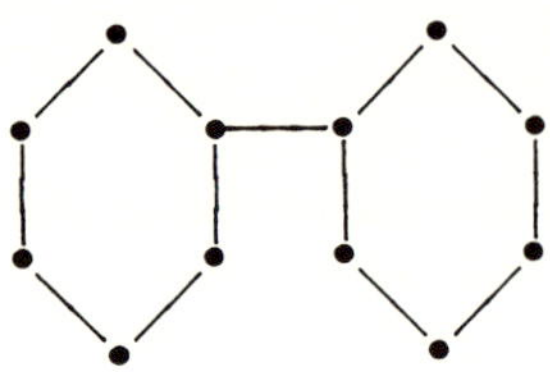			X

continued next page

Table 4—continued

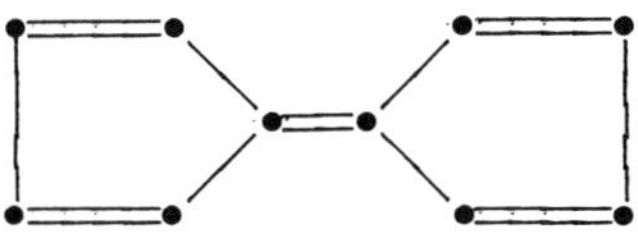		X
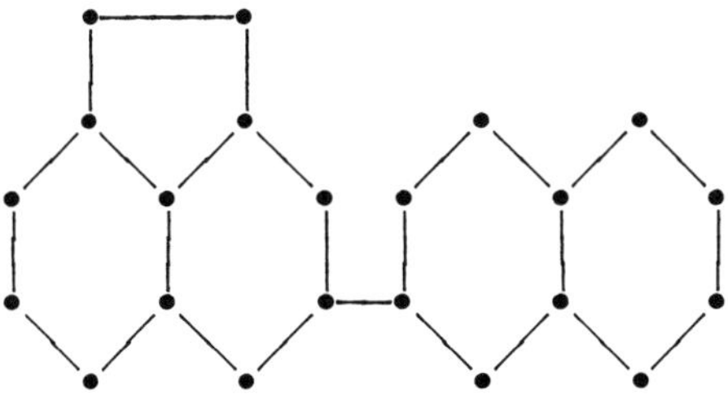	X	X

ing a notation for *ring aggregates,* the system of rings must be viewed in such a way that the system exhibits the *smallest set of smallest rings.* This requirement, necessary for the generation of unique notations, is nevertheless a very difficult one, especially when the ring aggregate is quite, but not perfectly, symmetrical. The derivation of notations for ring aggregates is complicated further by the necessity for assigning a unique numbering pattern to each ring aggregate, as will be seen subsequently. Examples 80 and 81 illustrate the notion of "smallest set of smallest rings," while examples 82 and 83 show the typical complications that can arise in choosing the proper set of rings.

	System	Preferred Set of Rings
80.	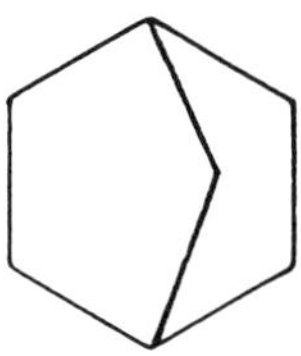	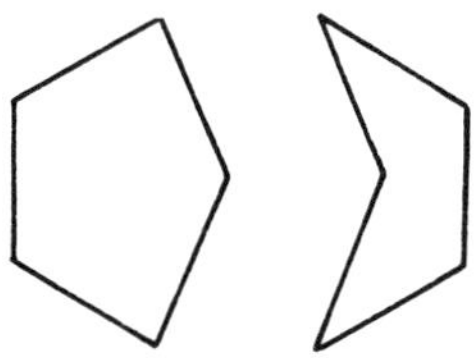

81. 

82. 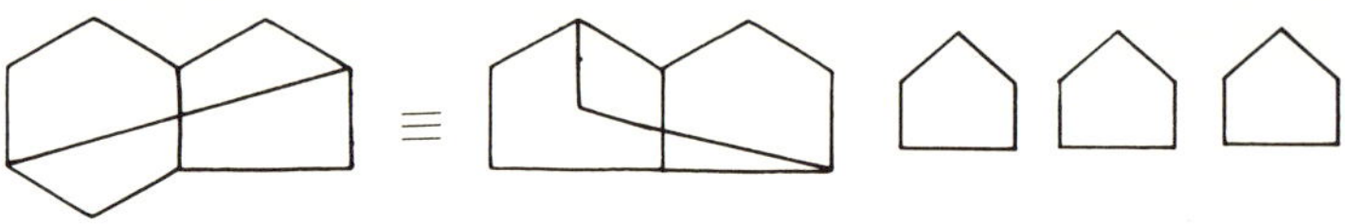

83. 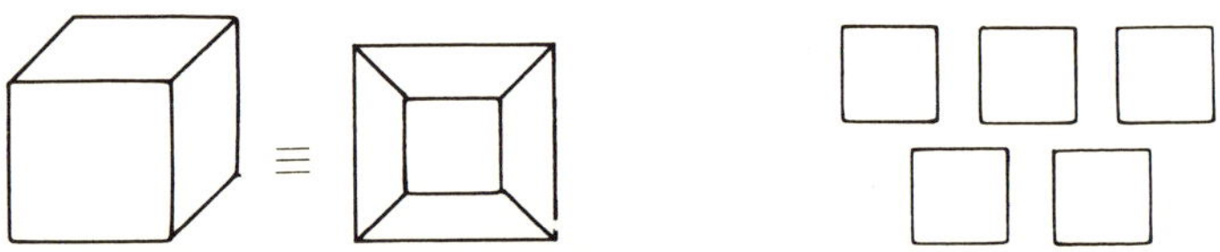

The principle upon which the numbering of the atoms in ring aggregates is based is not easily stated. In general, that ring is chosen in which to begin the numbering which has the greatest number of shared edges and/or atoms. When alternatives exist, that ring is chosen which will assure lowest locants for atoms shared by the several rings. Locant pairs are cited in the notation, in ascending numerical sequence, to indicate the "ring junctions." The following examples illustrate the numbering of ring aggregates:

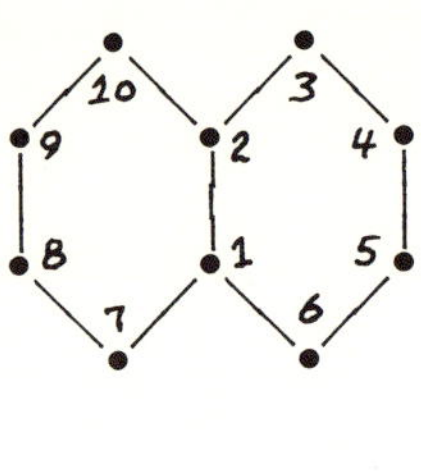

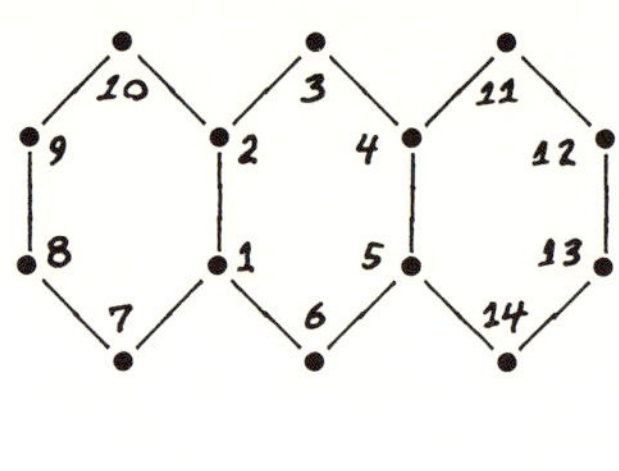

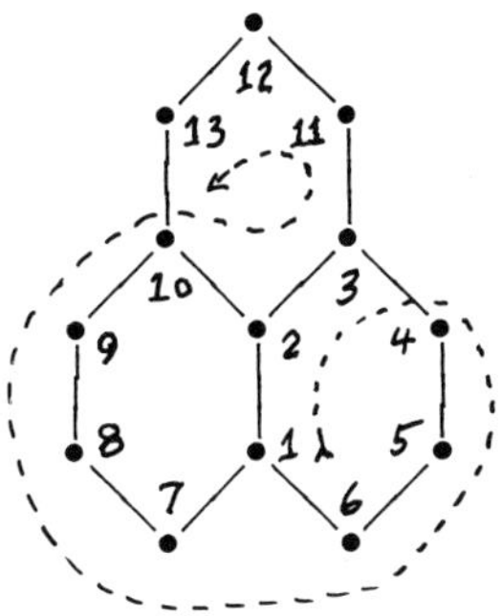

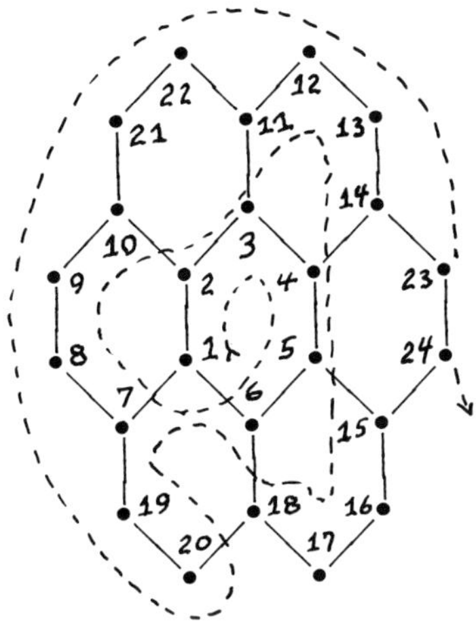

Notations for some typical structures are illustrated in the following examples:

84. 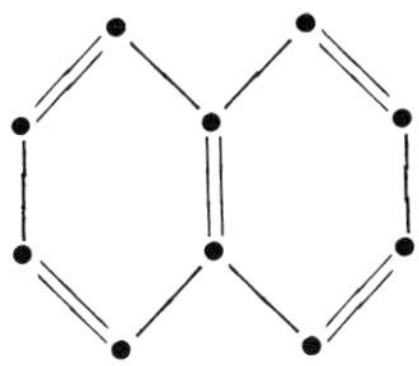 $B6_2$ *

85. 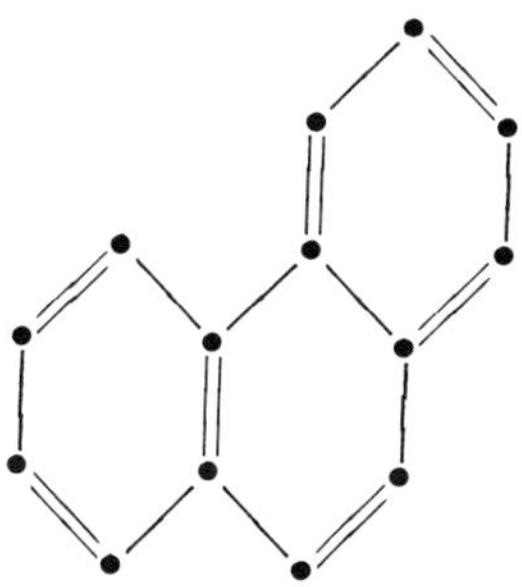 $B6_3$ 13**

*Since the ring juncture locants are 1-2 and since only the 1 would be cited, it is omitted.

**The ring juncture pairs are *1*-2 and *3*-4.

86. 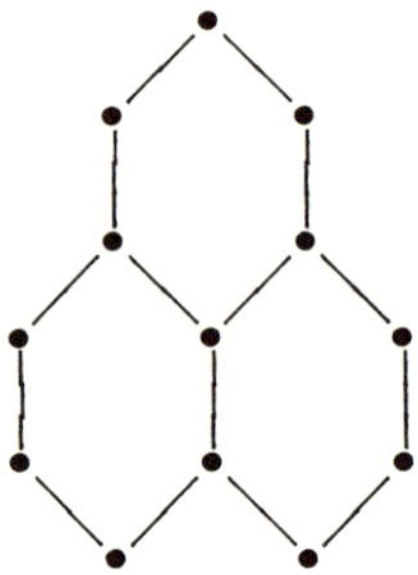 $A6_3$ 13-$\underline{10}$*

87. 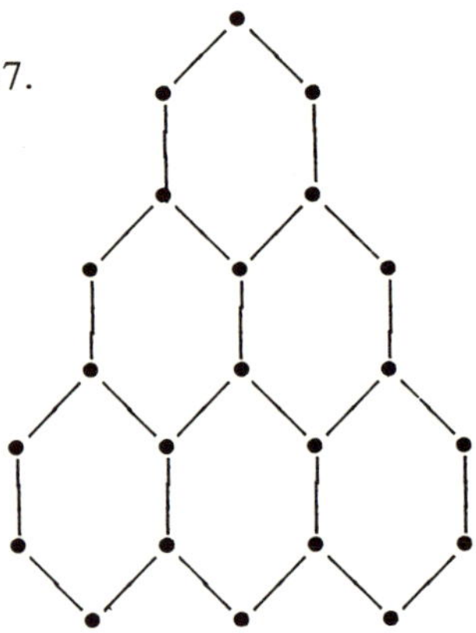 $A6_6$ 135-$\underline{141012}$-$\underline{19}$

Although we have by no means exhausted the types of compounds which the IUPAC notation system is capable of representing, and while we have left out many aspects of those types which were discussed, the material presented should give the reader a good appreciation of the IUPAC notation system. More detailed discussions of the system may be found elsewhere.[11]

The Wiswesser Line-Formula Chemical Notation

The linear notation system called the Wiswesser Line-Formula Chemical Notation and popularly referred to by the acronym "WLN" had

*3-10 since atoms 3 and 10 are not adjacent physically or numerically.

its first publication in 1953.[12] We noted earlier the interesting similarity between the symbolism employed in the original Dyson system and that employed by Wiswesser in the development of his system. But there ends the similarity between these two systems.* Whereas the Dyson system places first emphasis on the carbon skeleton of a compound, the Wiswesser system confers this emphasis upon functional groups. Nevertheless, the basic notion of describing chains of atoms, rather than branched aggregates, is basic to both the IUPAC and the Wiswesser systems.

Following the introduction of the Wiswesser system in 1954, it was adopted for use by several industrial firms. From the experience gained through its use, the Wiswesser system underwent revision, under the watchful eye of the Chemical Notation Association, whose members consisted of representatives of organizations experienced with the system. E. G. Smith was chiefly responsible for the revision of the notation system; the definitive edition (1968) of the rules for deriving and interpreting Wiswesser notations bears his name.[13] All illustrative material presented in this section is based on this source.

In a general way, the Wiswesser linear notation system can be said to assign a linear sequence of notation symbols to each linear sequence of atoms in a molecule. Consider the following structural formula:

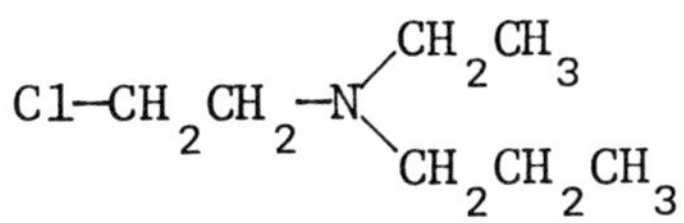

This formula may be thought of as consisting of two chains:

$$Cl{-}CH_2CH_2{-}N{-}CH_2CH_2CH_3 \qquad \text{and} \qquad {-}CH_2CH_3$$

The Wiswesser notation assigned to this compound will reflect this chain partitioning of the structure. (So, in fact, do many other linear notation systems, including the IUPAC system, but with differences

*Except, of course, for the fact that they both purport to represent the same compounds.

which will become clear.) In fact, the user is admonished to redraw the original structure as a two-dimensional diagram using notation symbols instead of standard element symbols.[14] Such a diagram is called a *graphic formula.* If we refer again to the above structure, and use the symbols G (for chlorine), N (for triply substituted nitrogen, no substituent a hydrogen atom) and the numerals 2 and 3 (for the carbon chains), we may write the "graphic formula"

$$\mathrm{G2N}^{2}_{3}$$

From this graphic formula it is very easy to derive the correct* notation for the compound in question. The derivation follows these basic steps. Choose the linear sequence of notation symbols which contain the greatest number of branch symbols (N, X, Y, K) and then the greatest number of notation symbols. Write this sequence from left to right beginning with the end symbol which is latest lexicographically.** We thus obtain G2N(2 or 3). The choice between 2 and 3 for the honor of appearing in the main sequence is also made on the basis of "lateness" in lexicographic order, so that the main sequence is G2N3. The remaining branch is cited in the notation following an ampersand (&), yielding the final notation, G2N3&2. With this introduction to the Wiswesser linear notation system, we turn to a more rigorous and detailed discussion of the system. In order to facilitate comparison of the systems discussed in this chapter, compounds used as numbered examples in the discussion of the IUPAC system will be used wherever practicable for illustration in the discussion of all the other systems. Wherever prac-

*We must use care in employing such terminology. While both the IUPAC and Wiswesser systems (among others) claim that unique and unambiguous notations are generated, the precise scope of these systems is unkown, and no mathematical proof (an exercise to be entered into only by the very zealous and hardy) exists to support such claims. However, to the extent that no counter example has, to our knowledge, been put forward to challenge the claim, one may say it is (so far) operationally proved.

**Lexicographic order is defined, in the Wiswesser system, as ƀ (blank) R & - / 0 1 2 3 4 5 6 7 8 9 10 11 12 . . . A B C D E F G . . . X Y Z *. "Latest" means nearest the right end of this list; thus ƀ is "earliest" and * is "latest."

ticable, the format for the discussion of these systems will parallel that used for the IUPAC system.

Acyclic Structures

Acyclic hydrocarbons are quite easily assigned notations in the Wiswesser linear notation system. Recalling the general discussion above, one first writes a graphic formula for the structure, then chooses for the main chain that sequence of notation symbols which has the greatest number of branch symbols, then the greatest number of symbols. Branches are cited in ascending order,* each branch being separated from the one before it in the notation by an ampersand. For example:

	Structure	Wiswesser Notation
9.	CH_4	1H
10.	$CH_3CH_2CH_3$	3H
11.	$CH_3(CH_2)_{18}CH_3$	20H
12.	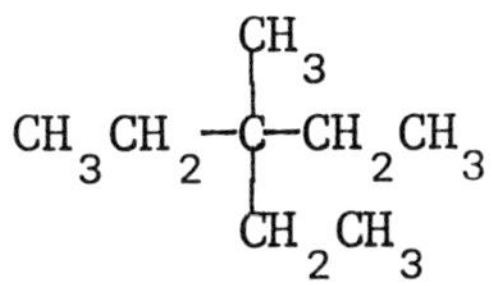	2X2&2**
13.	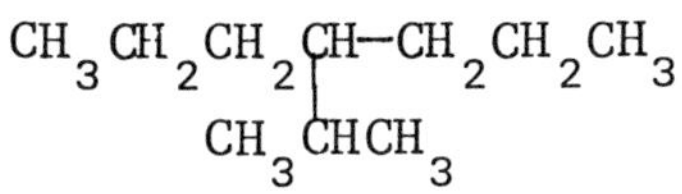	3Y3&Y

*Order is based first on that branch with the fewest branch symbols, then that branch with the fewest symbols, then the branch which is latest lexicographically.

**The "methyl contraction" principle obviates the necessity of citing the "1" in the notation representing the methyl group at C-3. Methyl groups are treated similarly in examples 13, 14, and 15.

14. 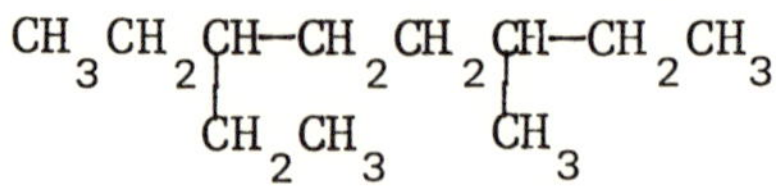 2Y2&2Y2

15. 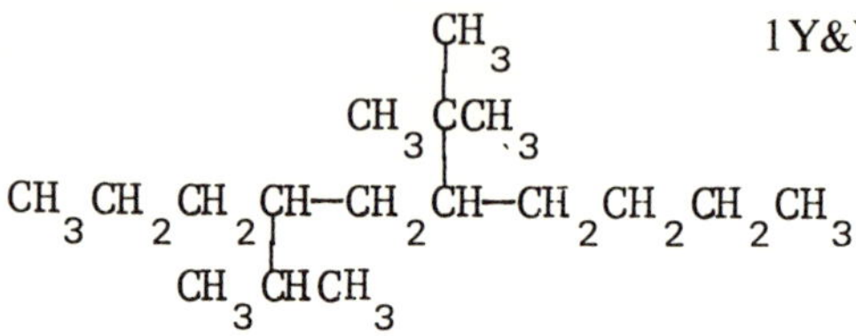 1Y&Y3&1Y4&X

Unsaturation in acyclic structures is generally denoted by the symbol U for double bonds, and UU for triple bonds, but certain exceptional cases must be noted, among which are carbon dioxide, ketenes, cyanates, isocycanates, and similarly structured compounds. The following examples illustrate the use of the U symbol:

16. $CH_3CH_2CH_2CH_2CH_3$ 5H

17. $CH{\equiv}CCHCH_2CH_3$ 4UU1

18. $CH_2{=}CHCH_2CH_2CH_3$ 4U1

19. $CH{\equiv}CCH_2C{\equiv}CH$ 1UU3UU1

20. $CH_2{=}CHCH_2C{\equiv}CH$ 1UU3U1

21. $CH_2{=}CHCH_2CH{=}CH_2$ 1U3U1

22. 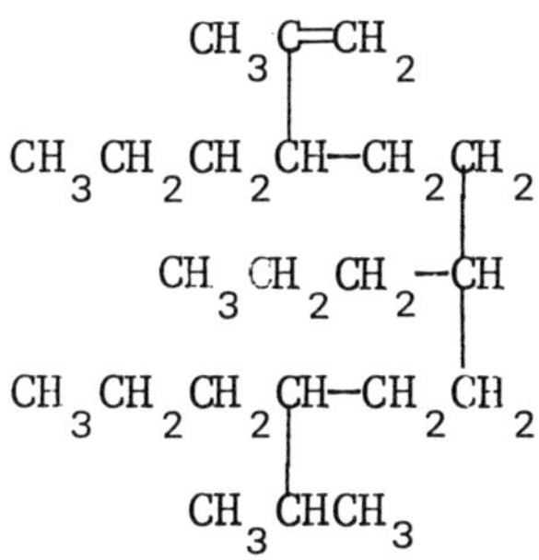 1Y&Y3&2Y3&2Y3&YU1

The following are examples of compounds having unsaturation *not* expressed by the U symbol:

88.	CO_2	OCO
89.	HC≡N	NCH*
90.	HON=O	QNO
91.	CH_2=N≡N	NNU1
92.	CH_3CH_2–O–C≡N	NCO2
93.	$CH_3CH_2CH_2$N=C=O	OCN3
94.	CH_3CH=C=O	OC2

As in the IUPAC linear notation system, the WLN system employs two methods of *contraction* to reduce the number of symbols which must be cited in a notation. The first of these is the "methyl contraction," which has already been illustrated in examples 12–15. The second method of contraction involves the use of *multiplicatives.* When

*Hydrogen (H) is cited here because the C symbol does not imply hydrogen attachment (i.e., C ≠ CH_4).

a graphic formula contains two or more identical chains of notation symbols attached to another symbol or symbol group, or attached to each other as object and mirror image, all the repeated symbols are cited once and the set of symbols is followed by a blank, and then by a numeral showing how many times the set appears in the structure. The use of multiplicatives is avoided unless four or more nonblank symbols are removed from the contracted notation. For example:

95. 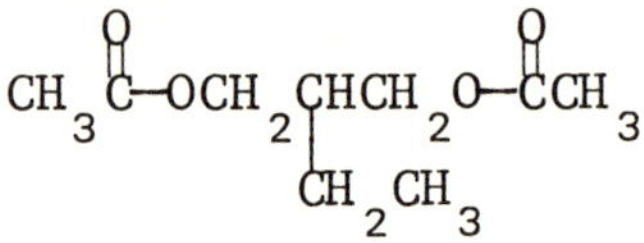 1VO1 2Y2

96. $(ClCH_2CH_2{-}S)_3CH$ G2S 3Y

97. 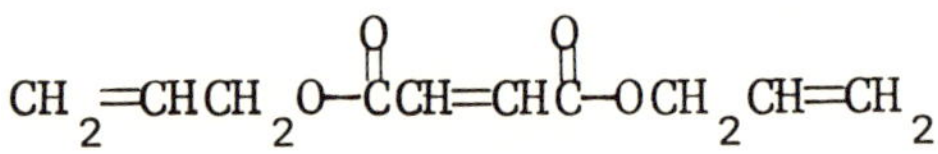 1U2OV1 2U

When the repeated set of notation symbols does not initiate a notation, a slash (/) is used to indicate the beginning of the repeated symbol set. For example:

98. 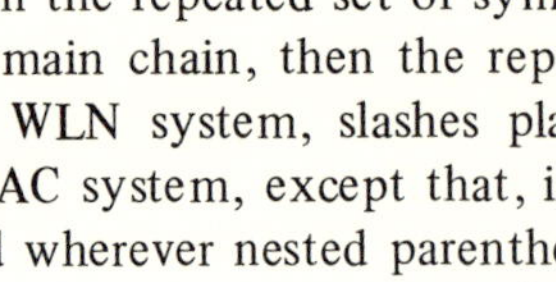 1VO1X/1O1 3

When the repeated set of symbols constitutes the main chain or part of the main chain, then the repeated unit is enclosed within slashes. (In the WLN system, slashes play the same role as parentheses in the IUPAC system, except that, in the latter system, assembly notation is used wherever nested parentheses would be required.) The multiplier is located in the notation as before. For example:

99. 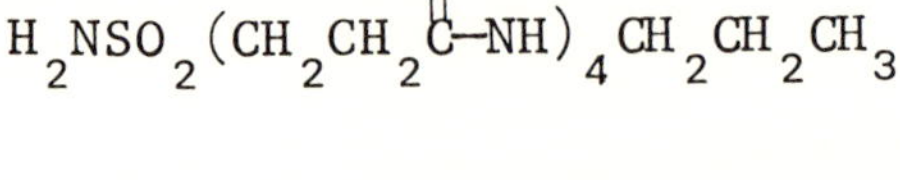 ZSW/2VM/ 43

In this discussion of acyclic structures, the main features of the Wiswesser linear notation system have been highlighted. Rigorous presentation of the rules has not been attempted; the reader should consult Smith's work for these details.

Cyclic Structures

In the Wiswesser linear notation system, benzene, among cyclic systems, is accorded special treatment. This treatment of benzene is based on the asumption that, because the benzene nucleus is so frequently found to be an element of organic structure, it merely "gets in the way" in indexing organic compounds and should therefore be subordinated and made as inconspicuous as possible. To this end, the benzene nucleus is represented by the symbol R. The R symbol carries six substitutable hydrogens, by implication. Thus:

62. 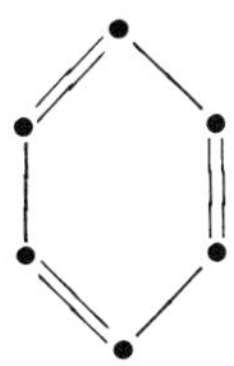

R

Benzene derivatives are categorized as mono-, di- or multisubstituted. In monosubstituted benzenes, the notation is written so that the R symbol comes last. For example:

100. 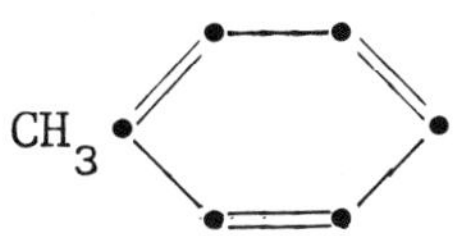

1R

101. 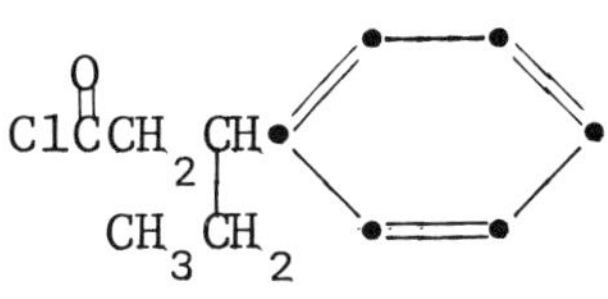

GV1Y2&R

102. 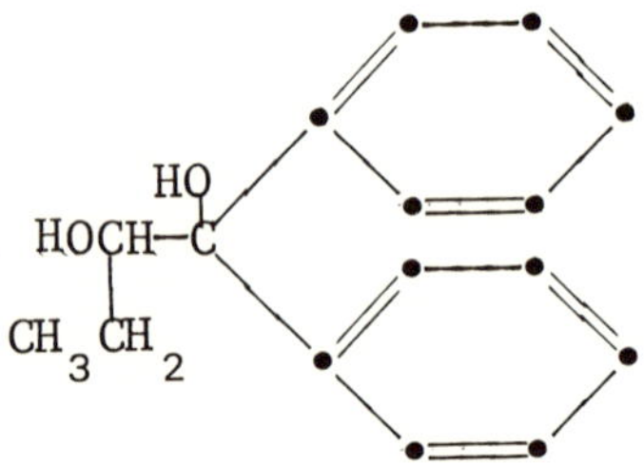

QY2&XQR&R

For disubstituted benzenes, the relative location of the substituents must be specified. The carbon atoms in benzene are numbered a, b, c, d, e, and f. The locant "a" is assigned to the part of the notation which precedes the R symbol; the other substituent is assigned the lowest locant possible consonant with the low numbering principle stated earlier. The following examples illustrate the treatment of disubstituted benzenes:

103. 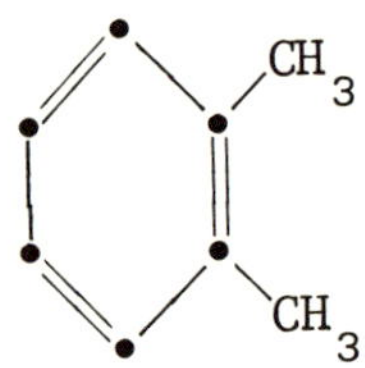

1R B

104. 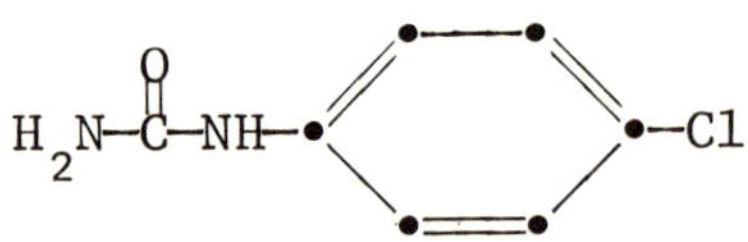

ZVMR DG

105. 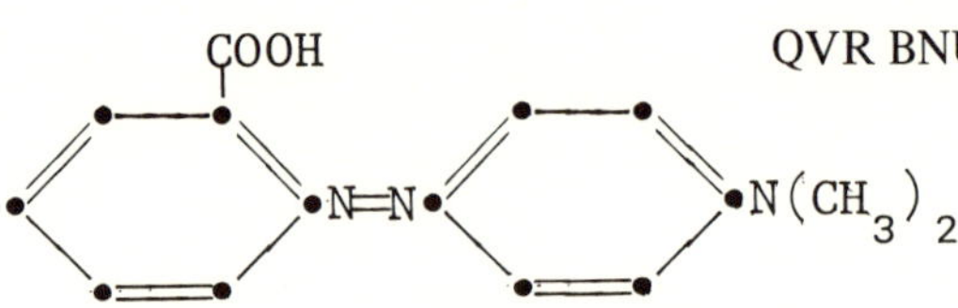

QVR BNUNR DN1&1

For multisubstituted benzenes, the additional task of choosing a main chain of notation symbols is imposed upon those already considered for mono- and disubstituted benzenes. In multisubstituted benzenes the R symbol is treated as a branch symbol to which the procedures already discussed in connection with acyclic compounds apply. Since locants are available, punctuation is normally not required to show where the substituents are located in the structure. In the following examples, note that the ring locants are the capital letters that immediately follow a blank in the notation.

106. 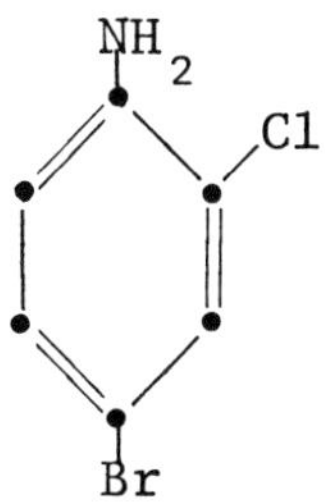

ZR BG DE

107. 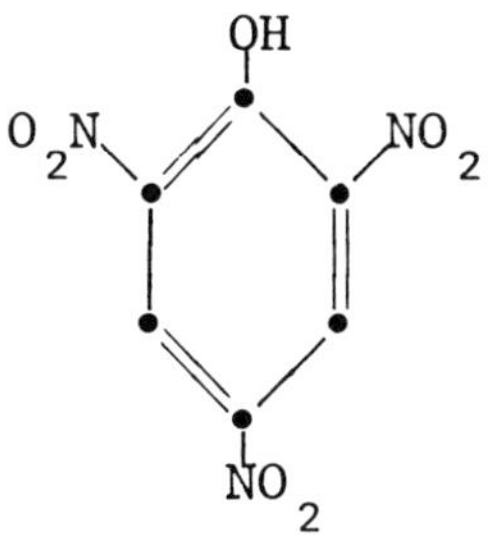

WNR BQ CNW ENW

108. 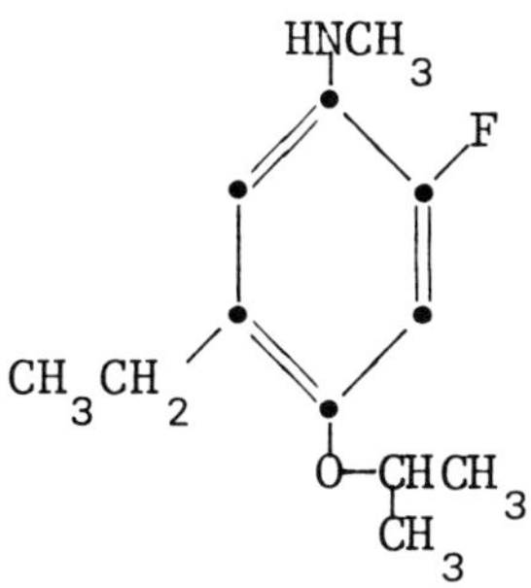

1Y&OR CF F2 DM1

109.* 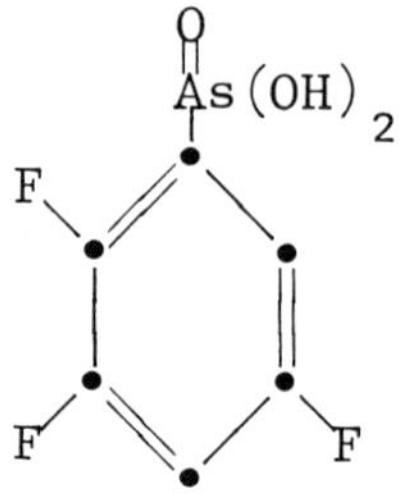 Q-AS-QO&R BF CF EF

Example 107 illustrates the low-numbering principle, as well as the principle concerning order of citation of branches in the notation. Example 108 admirably illustrates the principle for choosing the main chain of notation symbols (1Y&OR . . . . DM1). Other aspects of the treatment of benzene derivatives in the Wiswesser linear notation system include systematic contractions and benzene rings in branched chain structures. The reader is again referred to Smith for details.[15]

The Wiswesser linear notation system treats ring systems other than benzene. Carbocyclic ring systems are assigned notations that begin with the symbol L and end with the symbol J. Heterocyclic systems differ in this respect only in the starting symbol, which is T. Otherwise, the procedures for deriving notations for carbocycles and heterocycles are similar. Therefore both types of ring system will be treated together. In general, rings are considered to contain the maximum number of noncumulative double bonds, so that unsaturation is implicit in the notation. Ring atoms are given letter locants, as in benzene, and the locant path is chosen so as to yield low numbering. For example:

50.** 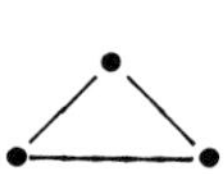 L3TJ

*Note that two-character atom symbols are surrounded by hyphens in the Wiswesser notation.

**If the ring is partly or fully saturated, a "T" symbol is used to indicate this fact. The symbol is placed immediately prior to the terminal "J" symbol.

51. 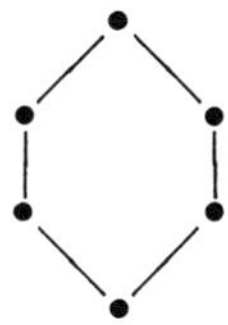 L6TJ

52. 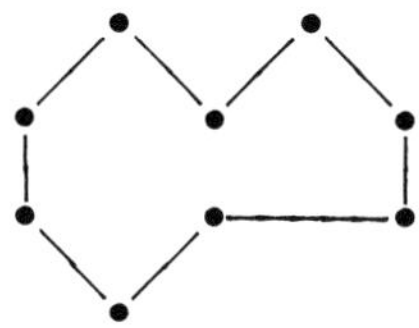 L9TJ

110. 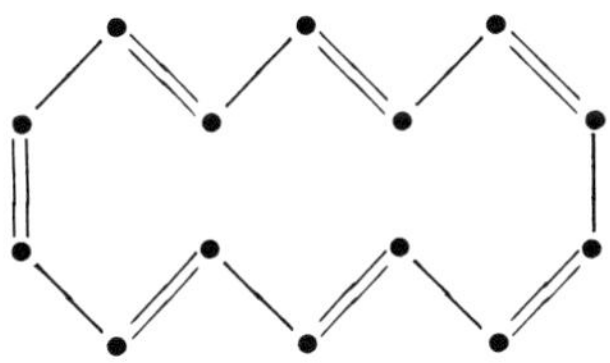 L-14-J

73. 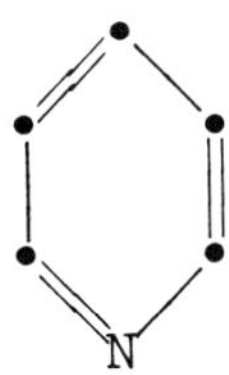

T6NJ

74. 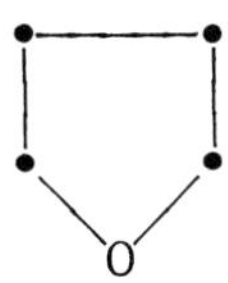

T5OTJ

75. 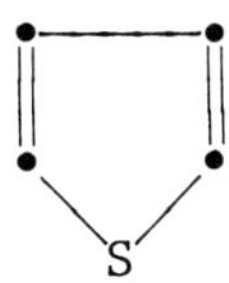

T5SJ

76. 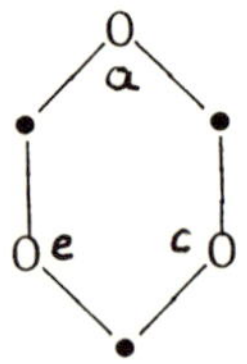

T60 CO EOTJ

111. 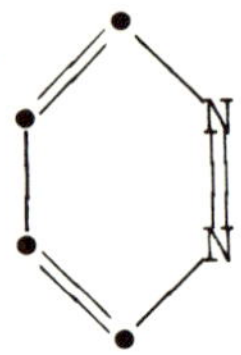

T6NNJ*

When the ring system under consideration has fewer than the maximum number of noncumulative double bonds, special procedures must be used. If the ring contains just a single carbon atom to which four other atoms are attached (not all four atoms in rings), the symbol H, together with the appropriate ring locant, is cited penultimate in the notation for the ring. If two or more such carbon atoms are present, the T symbol, without locants, is cited in the penultimate position in the notation for the ring. In this latter case, unsaturation which remains is made explicit by use of the U or UU symbols, which are placed in the notation immediately before any H or T symbols. For example:

65. 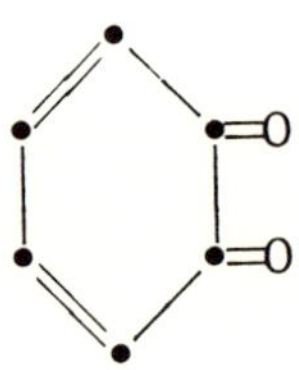

L6VVJ

*Consecutive locants are elided (here the locants for N and N would be A and B).

66. 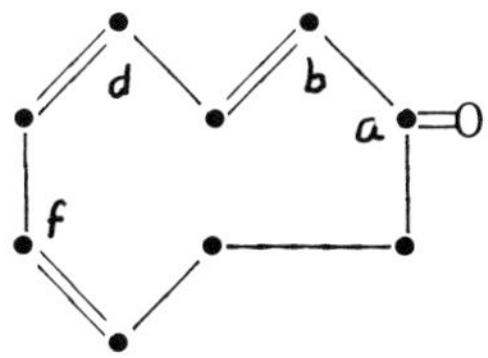

L9V BU DU FUTJ

67. 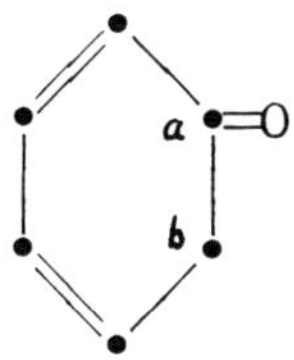

L6V BHT

112. 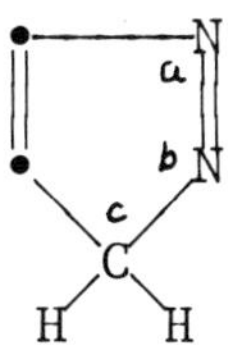

T5NN CHJ

113. 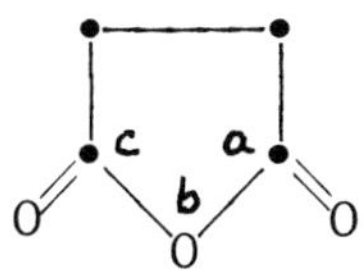

T5VOVTJ

Substituents on rings denoted by the L———J or T———J notation are handled in a manner similar to that for benzene derivatives, except that these ring notations are *not* treated as branch points as is the R symbol. We will not attempt to present here the rules for deriving notations for substituted ring systems, but merely give a few illustrative examples.

57. 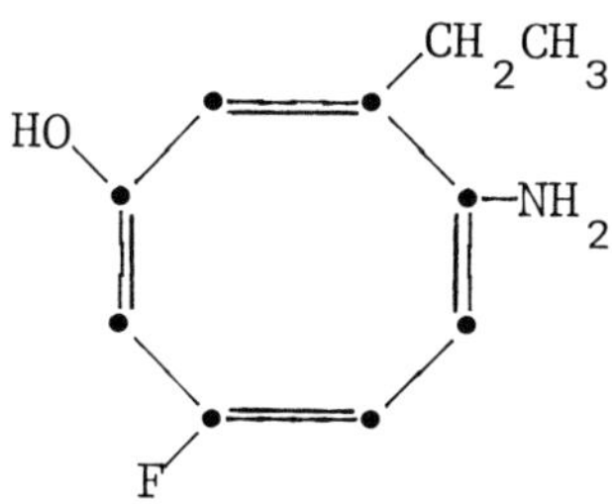

L8TJ AZ B2 DQ FF

59. 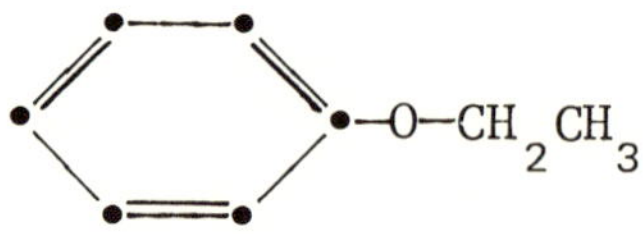 L6TJ AO2

114. 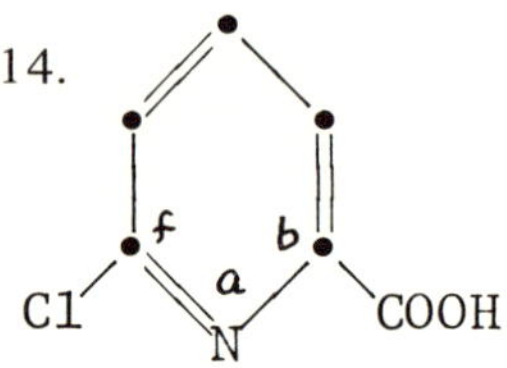 T6NJ BVQ FG

We come now to more complex ring systems. Because of the limited space available and because of the complexity of the procedures, only a superficial treatment of ring aggregates will be presented. The main problem to be dealt with in assigning any notation to a ring structure is that of labeling the nodes (atoms) of the ring(s) uniquely, for without a unique labeling, the description of the compound containing such rings can never be unambiguous. Most notation systems, among which nomenclature, the IUPAC linear notation system, and the Wiswesser linear notation system are important examples, seek a path through the nodes of the ring system which has a unique starting point, which traverses every atom of the system, and which doubles back on itself (that is, goes over old ground) as little as possible.

In the Wiswesser system, each node of the ring system is assigned an alphabetic locant (label), based on the following *partial* hierarchy of principles:[16]

Choose the locant path which does not cross a fused-ring junction (i.e., an edge [bond] shared by two rings), and which makes it possible in the following order of preference,

(1) to cite the smallest rings present,

(2) to cite the fewest rings necessary to fully describe the system,

(3) to have the fewest branch locants, all citable,

(4) to yield the lowest sum for fusion locants, and

(5) to yield the earliest set of fusion locants in the notation.

For complex ring systems, the procedure for defining a unique locant path obviously becomes quite difficult and is potentially very time-consuming. The following examples will illustrate the procedures. The reader who desires detailed procedures for assigning notations to ring systems is encouraged to refer to Smith.[17]

84.

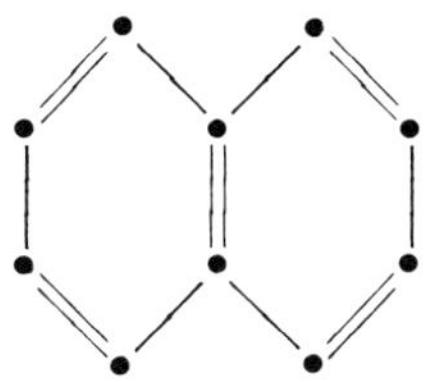

L66J

85.

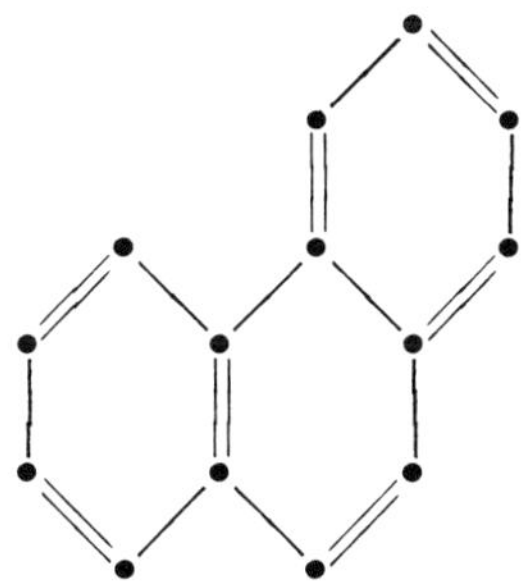

L B666J*

115.

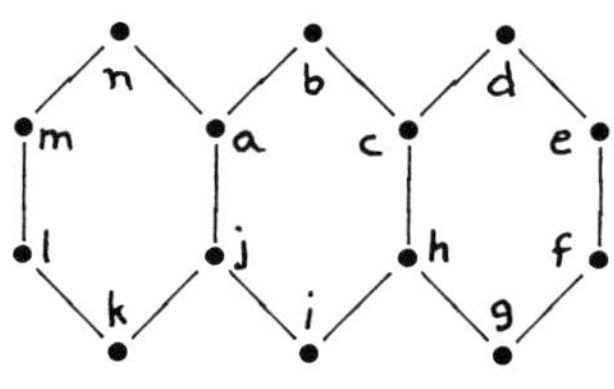

L C666TJ

*The lowest fusion locant in a ring is cited before its ring-size numeral, except "A" locants, which are omitted. Rings are cited earliest in the notation which are "completed" first by the locant path. Recall that locants are always preceded by a blank.

116.

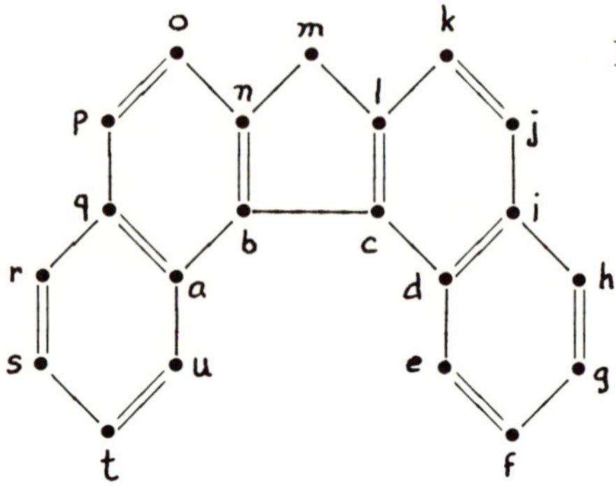

L D6 C6 B566 MHJ

86.

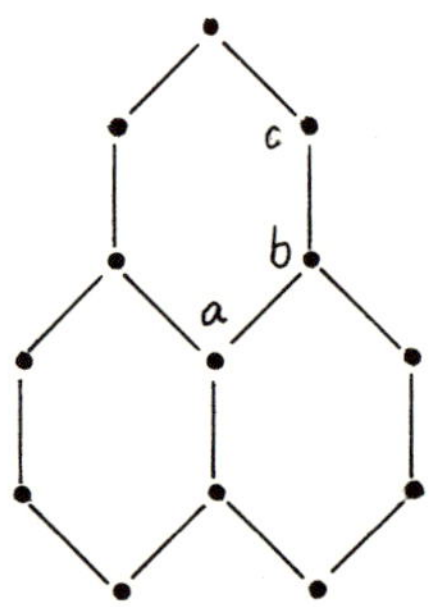

L666 1A MJ*

117.

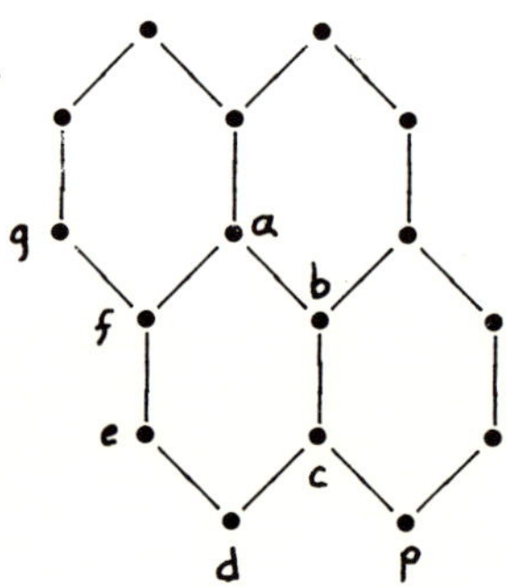

L666 B6 2AB PJ

*The ring sizes are cited in the order in which they are completed, then the number of, and locant for, each node shared by at least three rings (excluding spiro nodes), and finally the last ring locant in the ring system.

87. 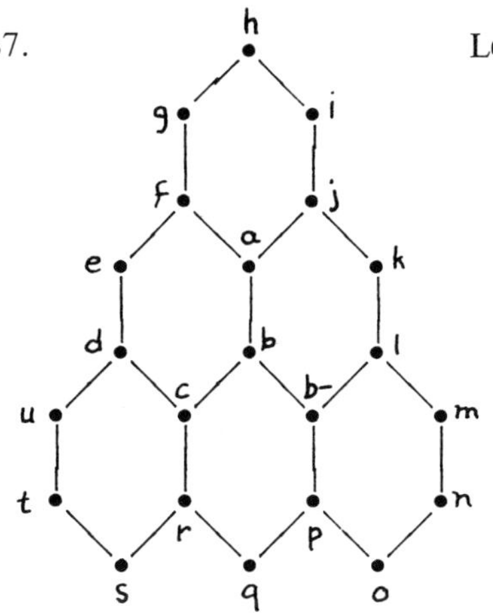

L666 B-6 B6 C6 4ABB-C U GH MHJ

The foregoing discussion of the Wiswesser linear notation system has touched upon those points which we believe are most commonly encountered in dealing with organic compounds. This discussion is by no means complete and we hope that the sketchy treatment given the various rules of the system does not lead to confusion or misunderstanding on the part of the reader. But the purpose of the discussion has been to highlight the principles of the Wiswesser linear notation system, not to make the reader an expert in the use of the system.

OTHER NOTATION SYSTEMS

The IUPAC and Wiswesser linear notation systems have been emphasized because of their importance in chemical documentation. In the remainder of this chapter several other notation systems will be discussed briefly. These are the Hayward, GREMAS, and Skolnik systems. A table of symbols and their significance in each of the several notation systems discussed is given at the end of the chapter. Of the notation systems to be described, the Hayward and the Skolnik systems purportedly generate unique and unambiguous linear notations. The GREMAS system yields unique, ambiguous linear notations (usually called fragmentation codes). Those systems which yield unique and unambiguous notations will be considered first.

The Hayward Linear Notation System

The Hayward linear notation system, developed as part of a research and development effort by the U.S. Patent Office, was first described in a report published in 1961.[18] The purpose of the notation system is to provide a unique and unambiguous representation of chemical structure which can be manipulated by computer to facilitate the task of the patent examiner in determining the patentability of an invention.

The Hayward linear notation system employs up to sixty-four different characters in forming notations for chemical species. The Hayward system differs from the IUPAC and Wiswesser systems in that it yields notations whose symbols are in one-to-one correspondence with the atoms and bonds of the molecule which the notation represents.* Like the IUPAC system, the Hayward system employs a saturated hydrocarbon skeleton as the basis for deriving notations. Let us first consider acyclic structures.

Acyclic Structures

The most important step in deriving a notation in the Hayward system (as in any other) is to properly enumerate the atoms of the structure. To do this, one considers a saturated hydrocarbon of the same constitution as the compound in question (exclusive of terminal atoms). For this hydrocarbon, the sequence of atoms "richest" in hydrogen is chosen as the main chain. Branches are enumerated starting with that branch attached to the lowest numbered atom in the main chain. Branches are numbered outward from the main chain. Suppose the following compound is to be represented in Hayward notation:

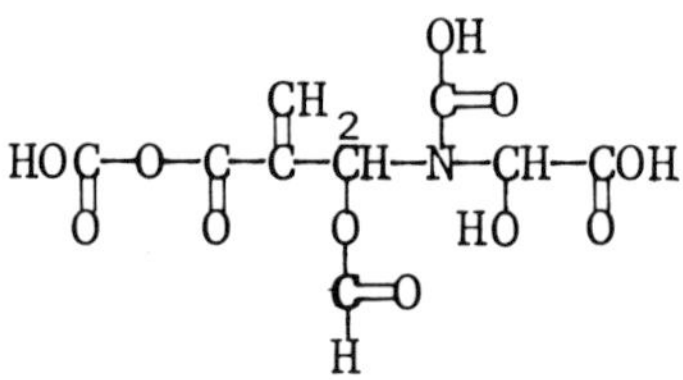

The corresponding saturated hydrocarbon is

*Exceptions are E, for "isolated" ethyl group; T, for "isolated" isopropyl group; W, for geminal dioxo (as in NO_2). Single bonds are not shown explicitly. "Isolated" means not joined to $-CH_2-$.

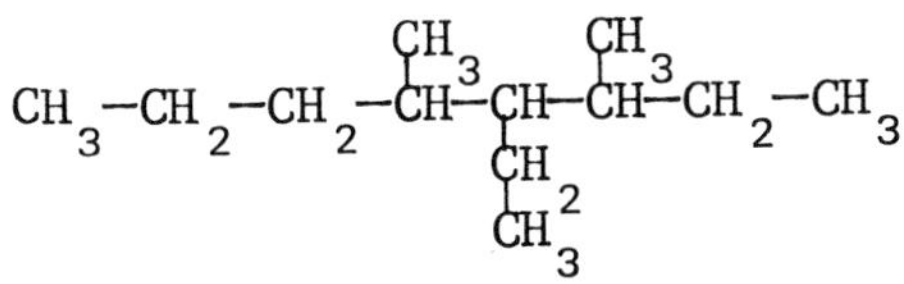

Since the path richest in hydrogen is

$$CH_3CH_2CH_2CHCHCHCH_2CH_3$$

enumeration begins with this atom sequence, starting from the left ($CH_3CH_2CH_2$- richer in hydrogen than -CH_2CH_3). The structure can now be numbered:

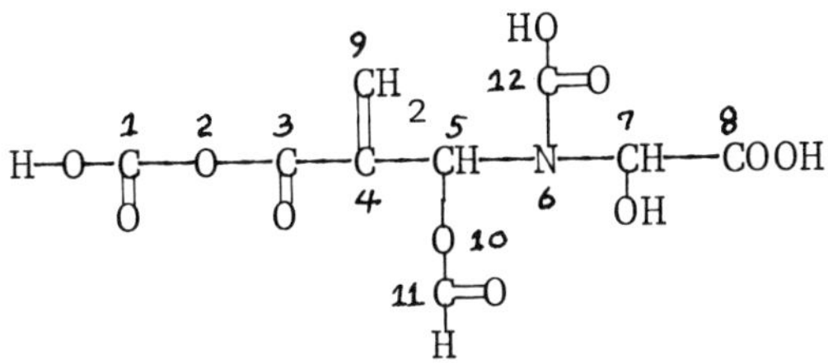

The notation for this structure is then formed as follows. (The main sequence atom numbers are shown above the notation.)

```
1   23 4     5     6      7 8
CVQOCVC(=C)C(OCV)N(CVQ)CQCVQ
```

Several examples will illustrate the notations derived from using the Hayward linear notation system.

Structure	**Hayward Notation**
12. 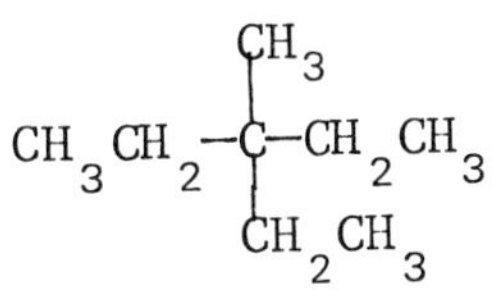	CCCME2

13.	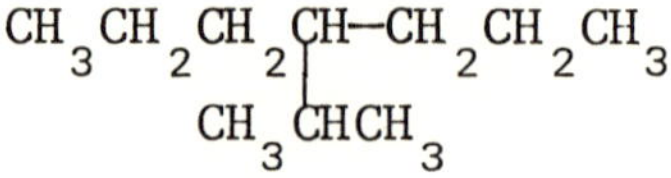	C3CTC3
17.	$CH{\equiv}CCHCH_2CH_3$	C÷CC3
18.	$CH_2{=}CHCH_2CH_2CH_3$	C=CC3
21.	$CH_2{=}CHCH_2CH{=}CH_2$	C=CCC=C
26.	$CH_3C{\equiv}N$	C(÷N)M
27.	$BrCH_2CH_2CH{=}NH$	CBCC(=N)*
28.	$CH_3CH_2CH_2NO_2$	CCCNW
29.	$CH_3(CH_2)_{10}CH_2SO_3H$	C11C(SWQ)
30.	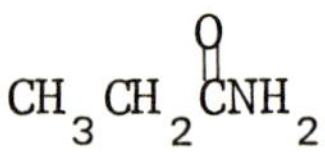	ECVZ
41.	$CH_3(CH_2)_4N{=}N(CH_2)_2CH_3$	C5N=NC3
118.	$(CH_3CH_2CH_2)_3N$	N(CCC)3
119.	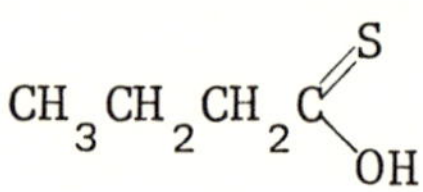	CQ(=S)C3

*B is used for bromine.

120. 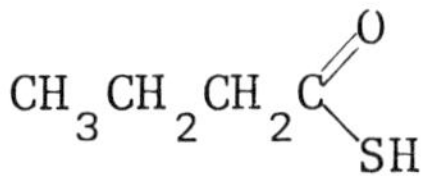

CV(SH)C3

Cyclic Structures

The Hayward linear notation system handles cyclic structures in the same manner as acyclic structures–on an atom-by-atom basis. Carbon atoms in rings are denoted, as appropriate, by the symbols Y for peripheral fusion atoms, L for carbon atoms in a "nonaromatic" ring,* and R for carbon atoms in an "aromatic" ring. Lower case *n* is used to represent ring nitrogen atoms. The enumeration of ring atoms follows the same principle as employed in the enumeration of acyclic structures. A notation begins with the size of the first and succeeding rings, if they are all of equal size; then the atoms of the ring, to the first fusion point, are stated. If the ring size of the next ring is different from that of the initial ring, the size of the new ring is cited after the ring fusion atom and the atoms in the path are cited as before. The following illustrates the method:

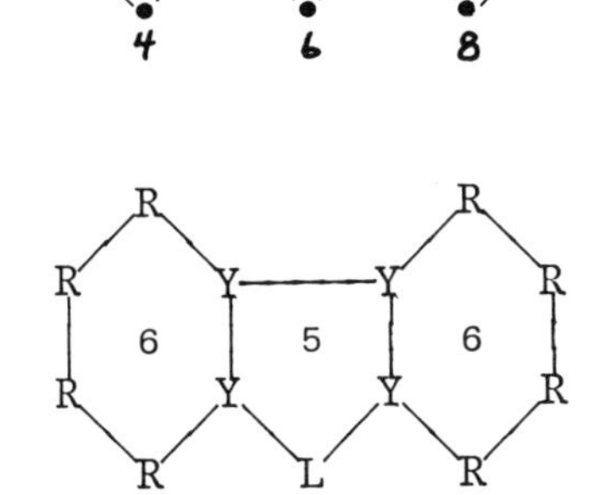

Step	Developing HLN
1	6
2	6RRRR
2′	6R4
3	6R4Y
4	6R4Y5
5	6R4Y5L
6	6R4Y5LY
7	6R4Y5LYRRRR
7′	6R4Y5LYR4
8	6R4Y5LYR4Y
9	6R4Y5LYR4YY

*I.e., a ring having fewer than the maximum number of noncumulative double bonds.

Substituents on rings are handled in the same manner as branches on acyclic structures. The following examples illustrate some of the provisions of the Hayward system for handling cyclic structures.

50. 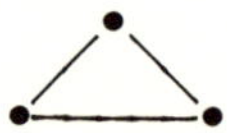 3L3

51. 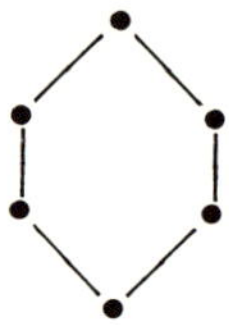 6L6

52. 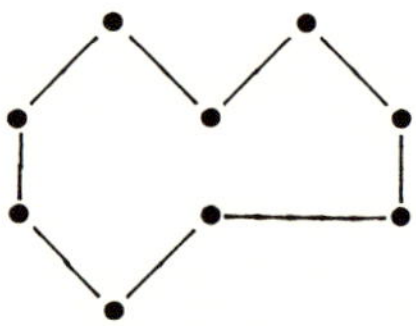 9L9

54. 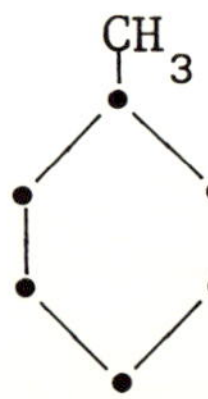

6LML5

55. 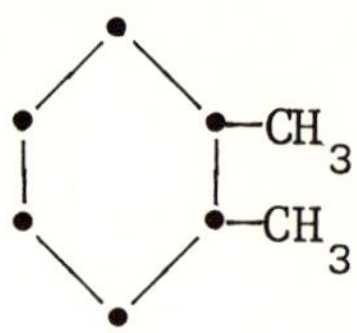

6LMLML4

60. 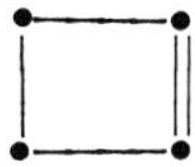

4L:LLL

62. 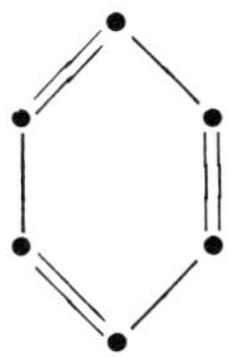

6R6

65. 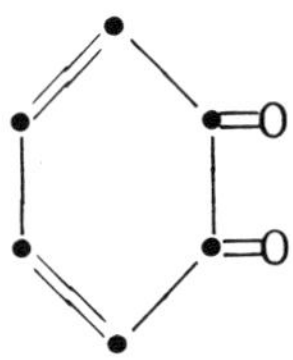

6LVLVL:LL:L

73. 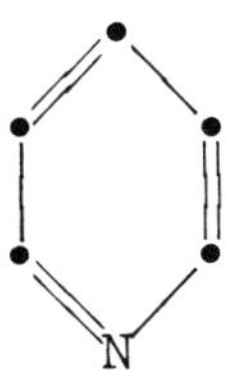

6nR5

121. 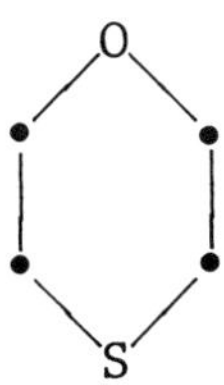

6SLLOLL

74.  5OL4

75. 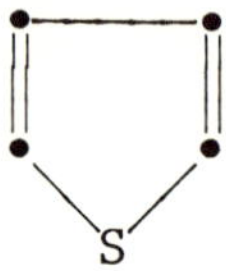 5SL:LL:L

79. 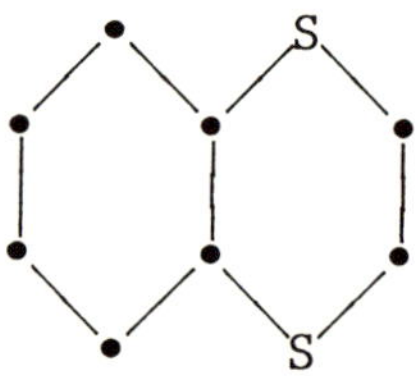 6SLLSYL4Y

84. 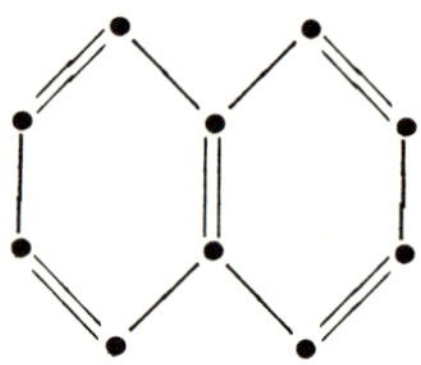 6R4YR4Y

85. 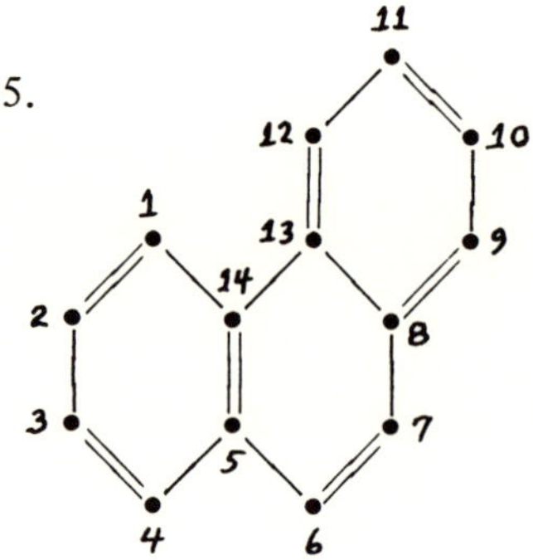

6R4YRRYR4YY

122. 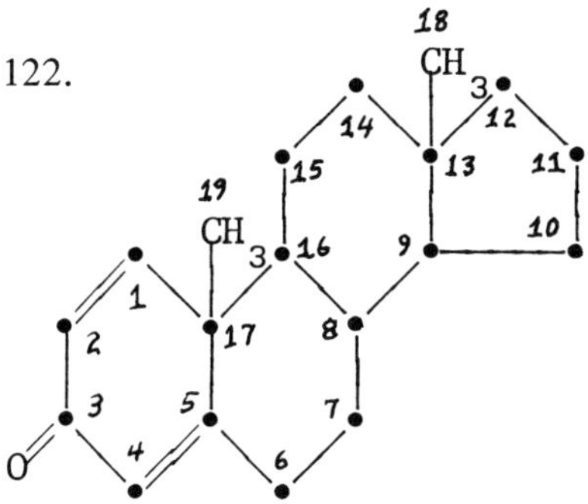 6L:LLVL:YLLYY5L3YMLLYYM

The Hayward system generally produces longer notations than does either the IUPAC or the Wiswesser system. On the other hand, the Hayward notations are more explicit representations of structure and, at least for relatively simple structures, are as easy to derive as are notations in either of the other systems so far discussed. The Hayward system has not, to our knowledge, been employed outside the Patent Office; some details of operational experience with the system have been reported.[19] Since it offers some interesting possibilities, it is perhaps regretable that it has received so little attention.

The Skolnik Notation System

Seldom has a linear notation system been devised with a specific aspect of information storage and retrieval as its goal. But such is the Skolnik linear notation system, a system which has been developed to facilitate the storage and retrieval of nuclear magnetic resonance (NMR) spectrographic data, as well as the correlation of such data.*

Skolnik announced a new linear notation system in 1969.[20] Subsequently, he described a modified version of that system for use with NMR data.[21] It is this latter system which will be described briefly here.

Since hydrogen nuclear magnetic resonance is most commonly

*Merely tying a structure to physical or biological data is not enough. Correlation implies the identification of a functional relationship between, say, structure and spectrum.

studied, it is obvious that a linear notation system which caters for NMR data must somehow make evident the atoms of hydrogen in a molecule. But it is also clear that, because hydrogen is the most commonly occurring element (in organic compounds, at least), a linear notation would become unwieldy if all hydrogen atoms were shown explicitly. A linear notation system must, therefore, exhibit hydrogen in a compact but easily interpretable manner. Skolnik accomplished this goal in a way analogous to that employed in the Hayward system–by providing symbols for each of the most common hydrogenated forms of carbon, as illustrated in Table 5. Heteroatomic groups are handled in a similar manner, as shown in Table 6.

The Skolnik linear notation system provides for the assignment of notations to structures in a manner parallel to that for the derivation of

Table 5–Skolnik Notation Symbols for Common Hydrogenated Forms of Carbon

Notation Symbol	Structural Significance of Notation Symbol	
	Structural Feature Indicated	Typical Structure
A	Methyl group	$CH_3CH_2\underline{CH_3}$
C	Methylene group	$CH_3CH_2\underline{CH_2}CH_3$
Y	Methylidyne group	$(CH_3)_3\underline{CH}$
J	Methylidyne group at bridgehead or ring fusion point	
E	Methene group	$CH_3CH{=}\underline{CH_2}$
B	Methenyl group	$CH_3\underline{CH}{=}CH_2$
U	Methyne group	$CH_3CH_2C{\equiv}\underline{CH}$
V	Methynyl group	$CH_3CH_2\underline{C}{\equiv}CH$
X	Methanetetrayl group	$(CH_3)_4\underline{C}$
D	Methenediyl group	$(CH_3)_2\underline{C}{=}CH_2$
R	Methenediyl group at ring fusion point	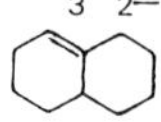
T	Methanetetrayl group at bridgehead, spiro union, or ring fusion point	
K	Carbonyl	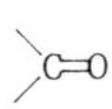

Table 6–Skolnik Notation Symbols for Common Heteroatomic Groups

Notation Symbol	Structural Significance of Notation Symbol	
	Structural Feature Indicated	Typical Structure
M	Secondary amine	$CH_3CH_2NHCH_3$
MH	Primary amine	CH_3NH_2
N	Tertiary amine	$(CH_3CH_2CH_2CH_2)_3N$
Z	Nitrile, azo	$CH_3C{\equiv}N$, $CH_3N{=}NCH_3$
Q	Ether, ester, peroxide	$CH_3CH_2OCH_2CH_3$
QH	Alcohol	$CH_3CH_2CH_2(CH_3)_2COH$
W	Non-linear "oxide"	$CH_3CH_2CH_2CH_2NO_2$
F*	Fluorine	$F_2C{=}CF_2$
G	Bromine	CH_3CH_2Br
L	Chlorine	$(CH_3)_3CCl$
I	Iodine	CH_3I

*Elements not shown are represented by their own element symbols using upper case letters. Two-character element symbols are preceded by an ampersand (&).

chemical names. (See Chapter 7 for details.) Thus, the structure is numbered starting from that end of a chain of atoms which has the highest priority. Branches are numbered in a similar fashion, but beginning at the point of attachment of the branch to the main chain. Parentheses are used in a notation to delineate branches attached to the main chain. Contraction of notations is limited to the use of multiplicatives. The following examples illustrate these principles:

Structure	Skolnik Notation
123. CH_3OH	HQA
124. 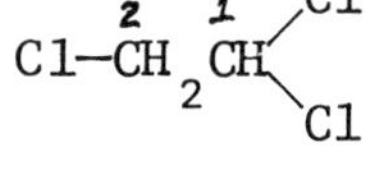	L2YCL

4. 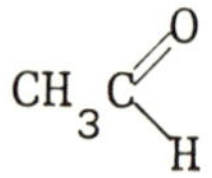 HKA

125. 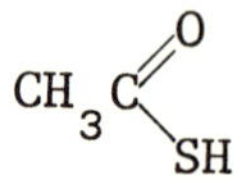 HSKA

126. $CH_3C{\equiv}CH$ UVA

127. 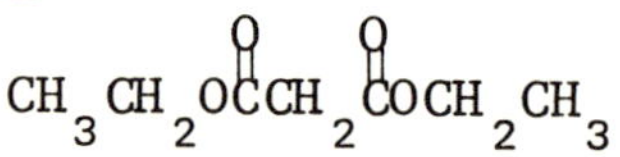 C(KQCA)2*

128. 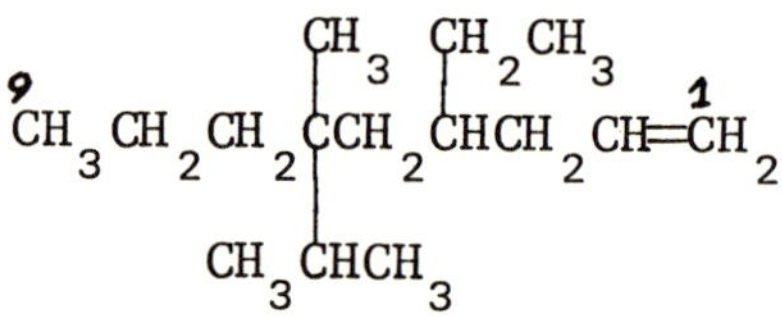 EBCY(CA)CX(A)(YA2)CCA**

In the Skolnik system, cyclic compounds are assigned notations in a manner analogous to that for acyclic structures (recall also the discussion of the Hayward linear notation system). Numbering of the atoms of a ring system follows Ring Index practice.[22] The lowest numbered ring atom is placed as early as possible in the notation and the entire sequence of ring symbols is set off by initial and terminal periods (.). Ring substituents are cited following the ring atom (notation symbol) to which they are attached. Atoms or atom sequences in bridges are preceded in the ring notation by a colon (:). If more than one bridge is present in the cyclic structure, these are cited in ascending order of bridgehead locants. A few examples will illustrate the handling of cyclic structures within the Skolnik system.

*Note contraction of notation by use of multiplicative "2" for repeated group, KQCA.

**Illustrates choice of main chain and numbering pattern.

50. 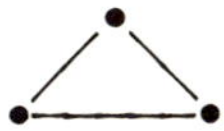 .CCC.

51. 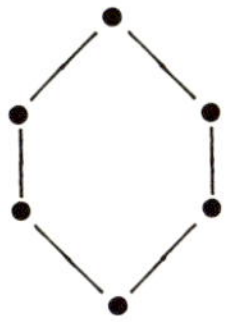 .CCCCCC. or .C6.

129.  .QCCC.

130. 

.MCCQCC.

131. 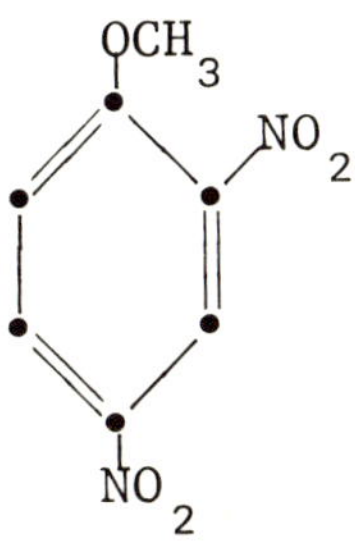

.D(QA)D(NW)BD(NW)BB.

132. 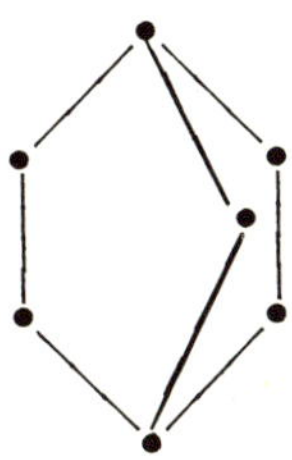 .JCCJCC:C.

133.

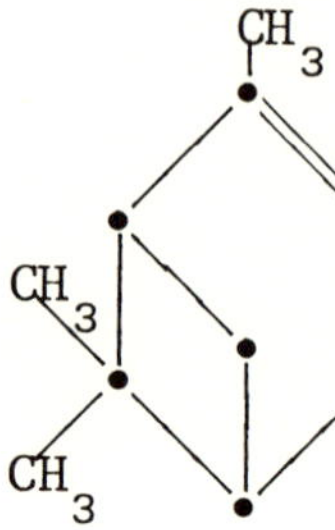

.JD(A)BCJX(A2):C.

84.

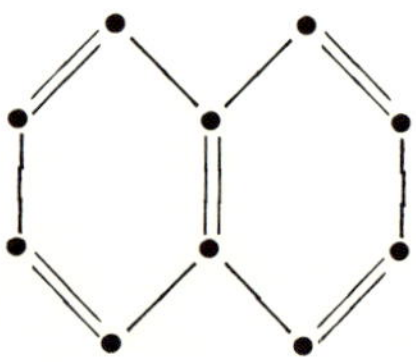

.BBBBR2□BBBB. or .B4R2□B4.

134.

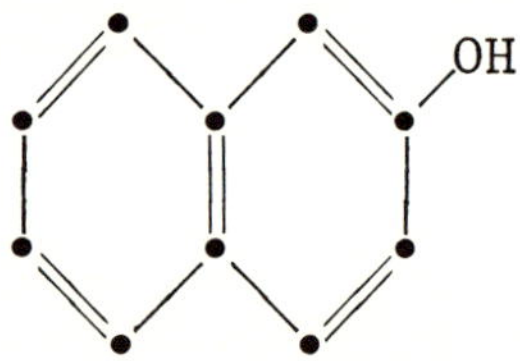

.BD(QH)B2R2□B4.

135.

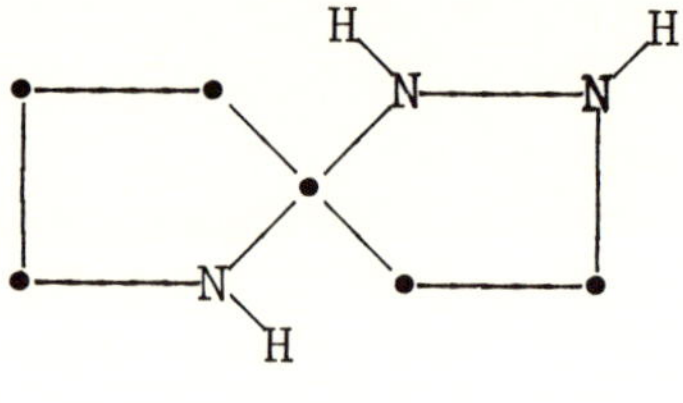

.MMCC□T□MCCC.

85. 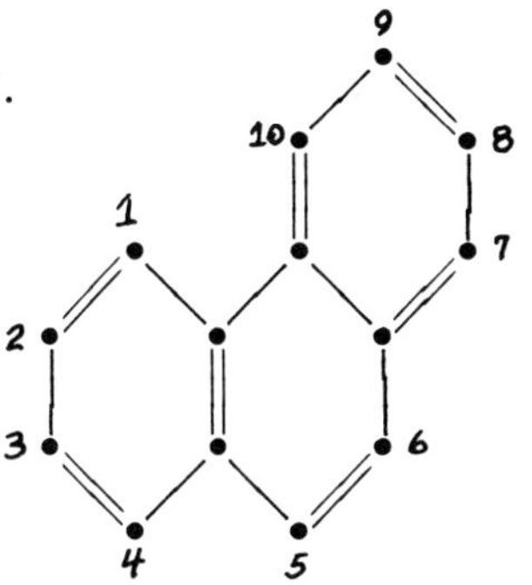 .B4R2□B2R2□B4.

Parallel indexes, based on the Skolnik notations and on the chemical shifts for each "proton group" represented in the notation, are easy to produce and can be useful tools for storing and retrieving NMR data.[23]

The GREMAS System

The last linear notation system which we shall consider is one which yields unique, ambiguous notations for chemical compounds. The system, called the *G*enealogisches *Re*cherchieren durch *Ma*gnetband-*s*peicherung (GREMAS) system, was first described by Fugmann, Braun, and Vaupel in 1961.[24] It was devised and is employed at Farbwerke Hoechst, A. G., Frankfurt/Main.[25] (A more recent discussion of the GREMAS system is that of Rössler and Kolb.)[26] In the GREMAS system, structures are represented by sets of three-letter symbols called *GREMAS terms.* Each GREMAS term corresponds to a certain fragment of a molecule. Among these are the individual carbon atoms (for which the condition of substitution is described), chains of carbon atoms, rings, and the linkages between heteroatoms. The system provides greatest differentiation among the more frequently occurring functional groups and elements. In addition to the GREMAS terms for individual atoms, terms are provided for ring structures, ring fusions, bridged atoms, unsaturation, chain length, and so on. Finally, terms are provided which allow specification of the approximate relationship between various structural features. An example will help to clarify these points. Consider oxydiacetic acid:

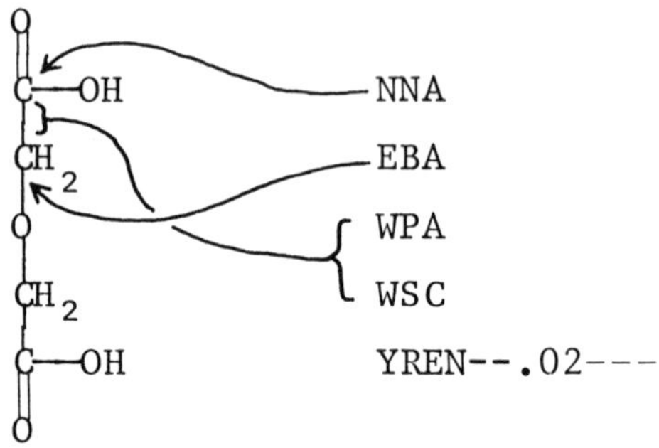

Since this structure has two identically similar carbon chains, the chain is described only once. The carbon of the carboxylic acid group belongs to *genus* N because the carbon atom has three bonds to heteroatoms. The genera N and O are common to all chain carbon atoms with "heteroorientation" III ("C^{III}"). Since both heteroatoms are oxygen, and since the singly bound oxygen is hydrogen-substituted or, at most, bears a negative charge, the carbon in question belongs to *species* N. Finally, because the carbon is attached to acyclic carbon by its one remaining bond, it belongs to *subspecies* A. Hence, the carboxylic carbon atom is assigned the GREMAS term NNA. The methylene carbon atom belongs to genus E because it has one bond to an oxygen atom; to species B because the oxygen is in an ether linkage; to subspecies A because its remaining bonds are to hydrogen or acyclic carbon. It is therefore assigned the GREMAS term EBA. The number of carbon atoms (2) contained in an (un-) branched chain is specified by the term WPA. Genus W signifies certain general structural features; species P limits the feature to the number of carbon atoms in an acyclic "skeleton"; A indicates that there are two such. WSC specifies that a carbon with one heteroatom attachment is linked to a carbon with three heteroatom attachments. Finally, a term beginning with genus Y is developed. Such a term is used to express which and how many substituents are present in an individual section of a molecule. In the GREMAS system, four typical sections are considered:

(1) chains of carbon atoms ("YR. . . .")

(2) alicycles ("YS. . . .")

(3) benzene rings ("YT. . . .")

(4) heterocycles ("YU. . . .")

In such "Y" terms all the substituents belonging to a certain section are represented by the genus-character of the corresponding carbon atoms. The "possible" substituents typically occurring in Markush formulas are enumerated following a stroke. Thus, they are distinguished from the essential substituents. Another device, which will not be discussed in detail, permits one to express whether an entire section is an essential part of a molecule or only a possible one and, in the latter case, by which other sections (if any) it can be replaced without changing the intended meaning of the whole Markush structure. The length of the Y terms is variable, depending on the size of the corresponding section. Blanks (here represented by dashes) are used to pad their length to an integral multiple of six. The number of occurrences of each section of a molecule is denoted by an appendix to the corresponding Y term, consisting of a period followed by the numerical frequency to be expressed. The complete set of terms is ordered alphabetically, so that we obtain for oxydiacetic acid the GREMAS notation: EBAN-NAWPAWSCYREN--.02---.

Before presenting other examples, we must pause to introduce additional GREMAS terms. Table 7 gives GREMAS terms for some of the more common functional groups. While this and later tables are by no means complete, they nevertheless indicate the scope and flexibility of the GREMAS system. Table 8 gives GREMAS terms for some chain features, such as unsaturation and length, and Tables 9, 10, and 11 present GREMAS terms for cycles and certain features of cycles. Armed with these tables, one should not find it difficult to comprehend the examples which follow. These examples are based, as far as practicable, upon those structures already employed in earlier discussions (2 in a YR term means $>C{=}C<$ and 3 signifies $-C{\equiv}C-$).

Structure	GREMAS Notation
1. $CH_3CH_2CH_2CH_2CH_2COOH$	NNA RAB RAC YRN---
2. $HOOCCH_2CH_2CH_2COOH$	NNA RAC WPD YRNN--
5. $CH_2{=}C{=}O$	IIF RBC WPA YRI2--
7. $HOCH_2CH_2CHO$	EAA LHA RAC WPB YREL--
8. $OHCCH_2CH_2COOH$	LHA NNA RAC WPC YRLN--
10. $CH_3CH_2CH_3$	RAB RAC WPB YR----
11. $CH_3(CH_2)_{18}CH_3$	RAB RAC WPT YR----
12. 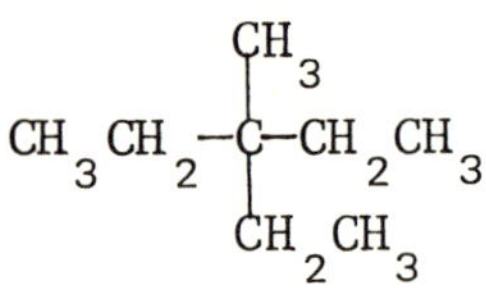	RAB RAC RAE WPG YR----
20. $CH_2{=}CHCH_2C{\equiv}CH$	RAC RBC RBD RCD RCE WPD YR23--
24. 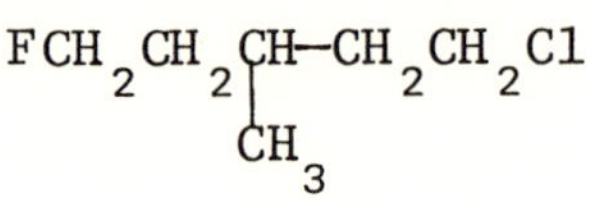	HAA HBA RAB RAC RAD WPE YRHH--
25. 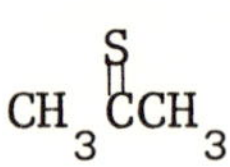	KIA RAB WPB YRK---

26. $CH_3C{\equiv}N$ — NBA RAB WPA YRN———

27. $BrCH_2CH_2CH{=}NH$ — HCA LBA RAC WPB YRHL——

28. $CH_3CH_2CH_2NO_2$ — CMA RAB RAC WPB XNF YRC———

29. $CH_3(CH_2)_{10}CH_2SO_3H$ — GAA RAB RAC WPX XSG YRG———

30. $CH_3CH_2\overset{O}{\overset{\|}{C}}NH_2$ — NGA RAB RAC WPB YRN———

31. $CH_3CH_2\overset{S}{\overset{\|}{C}}{-}Br$ — ORA O3A RAB RAC WPB YRO———

45. $CH_3CH_2CH_2\overset{O}{\overset{\|}{C}}{-}OCH_2CH_2CH_3$ — EDA NOA RAB RAC YRN——— WPB WPC YRE———

136. 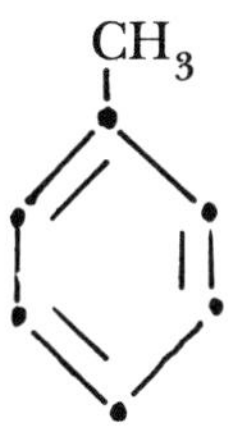

RAK SAF YRT——— YTR———

137 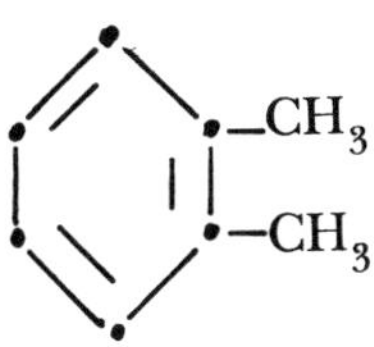

RAK SAF YTRR2–

57.

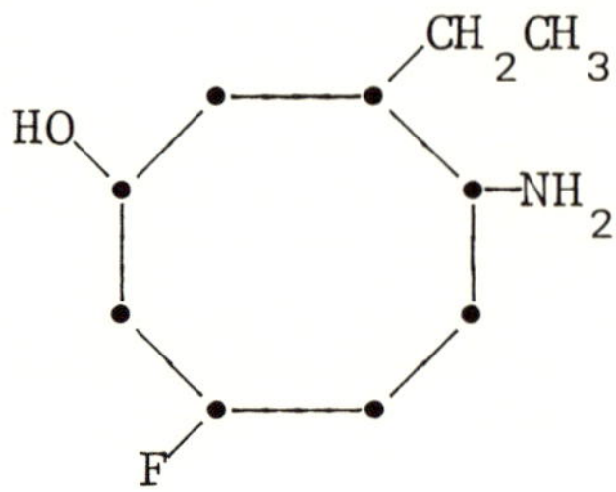

BAQ EAQ HAQ RAB RAL
SCL WPA YSBEH– YRS–––

63.

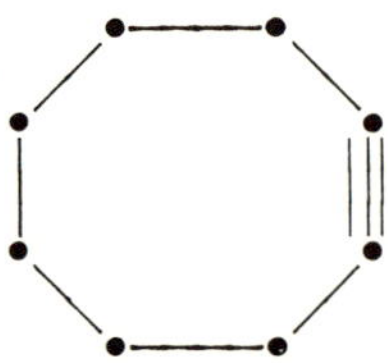

SCM TTC YS3–––

138.

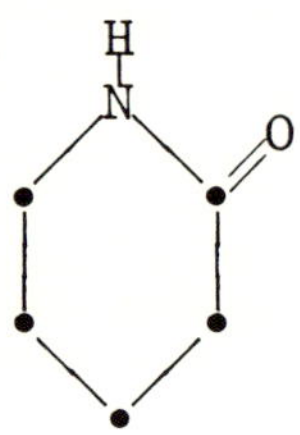

IIS SEO WHC YUI–––

139.

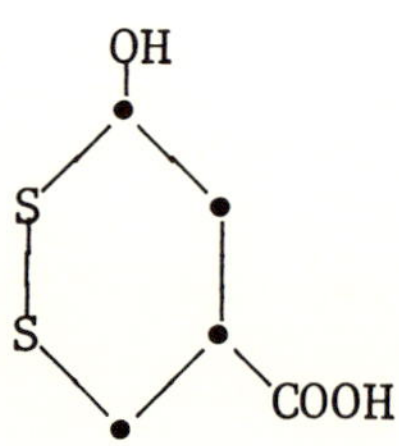

EAS NNE SOR WGA WHM
YUEN––

140.

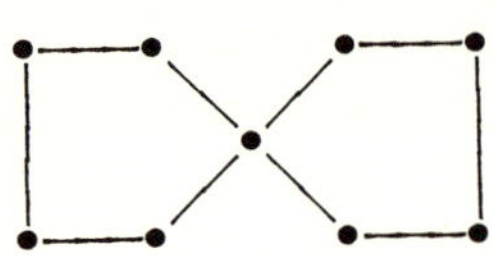

TCC TQA VAA YSV––– .02–––

141. 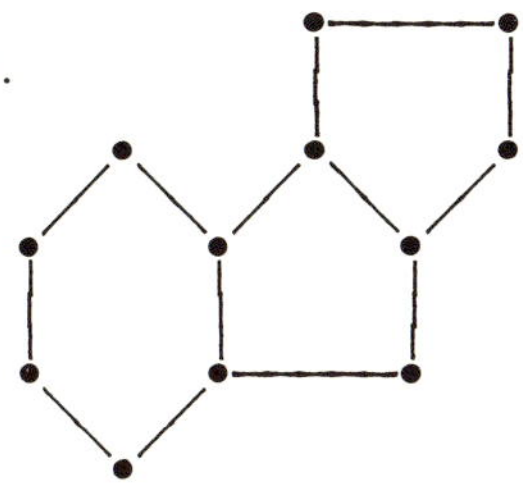

TAA TCC VBB VDB YTVV2–
YSVVVV YSVV––

142. 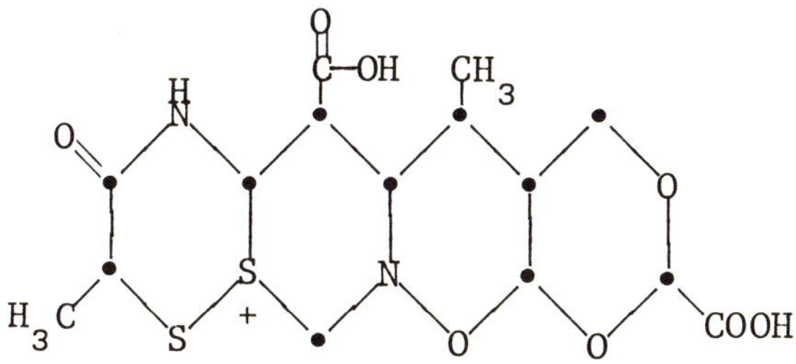

IIS NNE TGM TKD TMF TKO
RAO VGA VGB VGC VGL
VHB VHH VHL VHM WGA WGI
WGM WHC XH2 YUIRVV
YUNVVVV––––– YURY––
YRU––– .02–––
YURVVVV–––––
YUNVV–

These examples, though limited in number and scope, should give the reader an appreciation of the descriptive power of the GREMAS linear notation system. The GREMAS notation can be automatically generated from topological representations, a procedure carried out routinely by the Internationale Dokumentationsgesellschaft für Chemie m.b.H. (IDC).[27] This fact means that GREMAS notations also serve as efficient screens (classifications) for a file of topological representations of structure. Screens of this type have been established by IDC for individual, nongeneric compounds.

Table 7–GREMAS Terms for Some Common Functional Groups

GREMAS Term: Genus	Species	Subspecies	Structural Significance of GREMAS Term: Functional Group Class	Typical Structure (Atom represented is underscored)
A	B	A	Alkyl sodium	$CH_3CH_2CH_2\underline{CH_2}Na$
A	D	A	Alkyl magnesium	$CH_3\underline{CH_2}Mg-$
B	A	A	Primary amine	$C_5H_{11}\underline{CH_2}NH_2$
B	B	A	Secondary amine	$CH_3\underline{CH_2}-NH-CH_2CH_3$
C	A	A	Alkylidenamines	$CH_3\underline{CH_2}-N{=}CHCH_3$
C	C	A	Amido derivs.	$(CH_3)_3\underline{C}-NHC(=O)-CH_3$
C	E	A	Carbonamidic acids	$(CH_3)_2\underline{CH}-NHC(=O)-O-$
C	M	A	Nitro derivs.	$CH_3CH_2\underline{CH_2}NO_2$
D	A	A	Hydrazines	$CH_3-NH-NH-\underline{CH_3}$
D	B	A	Hydrazones	$CH_3\underline{CH_2}NHN{=}CHCH_2CH_2CH_3$
D	D	A	Hydrazides	$CH_3\underline{CH_2}-NHNH-C(=O)-CH_2CH_3$
E	A	A	Alcohols	$CH_3\underline{CH_2}OH$
E	B	A	Ethers	$CH_3O\underline{CH_3}$
E	D	A	Esters	$CH_3\underline{CH_2}O-C(=O)CH_2CH_2CH_3$
E	F	A	Carbonate esters	$CH_3\underline{CH_2}O-C(=O)-OCH_2CH_3$
F	-	-	Genus E with sulfur in place of oxygen	
G	A	A	Sulfonic acids	$CH_3(CH_2)_{10}\underline{CH_2}SO_3H$
G	C	A	Sulfoxides	$CH_3-S(=O)-\underline{CH_3}$
H	A	A	Fluorine derivs.	$CH_3CH_2CH_2\underline{CH_2}F$
H	B	A	Chlorine derivs.	$\underline{CH_3}Cl$
H	C	A	Bromine derivs.	$(CH_3)_2CH\underline{CH_2}Br$
H	D	A	Iodine derivs.	$CH_3\underline{CH_2}I$
I	I	A	Ketones	$CH_3CH_2\underline{C}(=O)CH_2CH_2CH_3$
I	K	A	Ketals, acyclic	$(CH_3)_2\underline{C}-(OCH_2CH_3)_2$
I	C	A	Imines, N acyclic	$CH_3\underline{CH}{=}NH$ (N not attached to N or O)
I	F	A	Oximes	$(CH_3)_2CHCH_2CH_2\underline{CH}{=}N-OH$
K	-	-	Genus I with sulfur in place of oxygen	

Table 7–continued

GREMAS Term: Genus	Species	Subspecies	Structural Significance of GREMAS Term: Functional Group Class	Typical Structure (Atom represented is underscored)
L	H	A	Aldehydes	$CH_3CH{=}CHCH_2\underline{C}H{=}O$
L	I	A	Acetals (& hemi)	$CH_3\underline{C}H{-}(OCH_2CH_3)_2$
L	D	A	Hydrazones	$CH_2\underline{C}H{=}N{-}NHCH_2CH_3$
L	E	A	Oximes	$CH_3(CH_2)_8\underline{C}H{=}N{-}OH$
M	-	-	Genus L with sulfur in place of oxygen	
N	B	A	Nitriles	$CH_3CH_2\underline{C}{\equiv}N$
N	C	A	Amidines	$CH_3CH_2CH_2CH_2\underline{C}({=}NH){-}NH_2$
N	G	A	Amides	$CH_3\underline{C}({=}O){-}NHCH_2CH_2CH_3$
N	L	A	Hydrazides	$CH_3CH_2\underline{C}({=}O){-}NH{-}NHCH_2CH_3$
O	-	-	Genus N with sulfur in place of oxygen	
P	A	A	Guanidine derivs.	$CH_3NH{-}\underline{C}({=}NH){-}NHCH_3$
P	D	A	Urea derivs.	$CH_3CH_2NH{-}\underline{C}({=}O){-}NH_2$
P	H	A	Carbonic acid der.	$CH_3O{-}\underline{C}({=}O){-}OCH_3$
P	P	A	Carbon dioxide	$O{=}\underline{C}{=}O$
Q	-	-	Genus P with sulfur in place of oxygen	

Table 8–GREMAS Terms for Some Common Features of Carbon Chains

GREMAS Term: Genus	Species	Subspecies	Remarks	Structural Significance of GREMAS Term: Structural Feature Indicated	Typical Structure (Feature Marked)
R	A	F	Generic terms	Saturated hydrocarbon	$CH_3CH_2CH_2CH_3$
R	B	F		Olefinic hydrocarbon	$CH_3CH{=}CHCH_3$
R	C	F		Acetylenic hydrocarbon	$CH_3C{\equiv}CCH_3$
R	A	B	Only carbon or hydrogen attached	Methyl group	$\underline{CH}_2CH_2CH_2OH$
R	A	C		Methylene group	$CH_3\underline{CH}_2CH_2SH$
R	A	D		Methylidyne group	$(CH_3)_2\underline{C}HCH_2Br$
R	A	E		Tetrasubstituted C	$(CH_3)_4\underline{C}$
R	B	C		Specific olefinic links	$H_2\underline{C}{=}\underline{C}H_2$
R	C	D		Specif. acetylenic links	$H\underline{C}{\equiv}\underline{C}H$
R	E	E		Allenes	$H_2C{=}C{=}CH_2$
W	P	A	2	Number of carbon atoms in a chain	$\underline{CH}_3\underline{C}OOH$
W	P	B	3		$\underline{CH}_3\underline{CH}_2\underline{CH}_2OH$
W	P	C	4		$\underline{CH}_2{=}\underline{CH}\underline{CH}_2\underline{CH}_2N(CH_3)_2$
W	P	D	5		.
W	P	E	6		.
W	P	F	7		.
W	P	G	8		.
W	P	H	9		.
W	P	I	10		.
W	P	K	11		.
W	P	L	12		.
W	P	M	13		.
W	P	N	14		$\underline{CH}_3\underline{C}H(H_3C)(\underline{CH}_2)_{10}\underline{C}{\equiv}\underline{C}O\overset{O}{\overset{\Vert}{C}}CH_3$
W	S	–	X = hetero attached	1,2 orientation of hetero substituted chain carbons	
W	T	–		1,3 orientation of hetero substituted chain carbons	
W	S	A		X–C:C–X	$NH_2\underline{CH}_2\underline{CH}_2NH_2$
W	S	E		$(X)_2C{:}C(X)_3$	$O{=}\underline{CH}\underline{C}({=}O){-}OH$

Table 9–GREMAS Terms for Some Simple Heterocycles

GREMAS Term	Typical Structure	GREMAS Term	Typical Structure
SDE		SNF	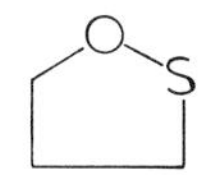
SLB	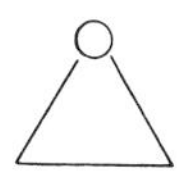	SEL	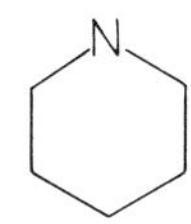
SDH	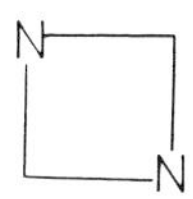	SEM	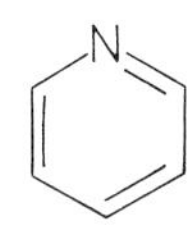
SIE	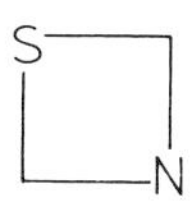	SEN	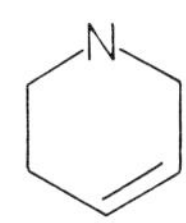
SEA	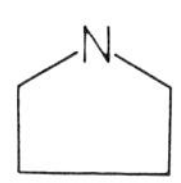	SEO	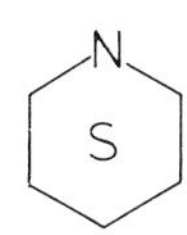
SEB	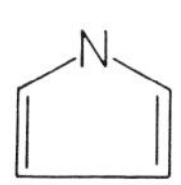	SKP	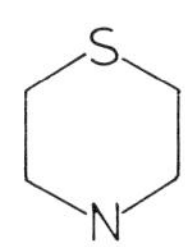
SHB	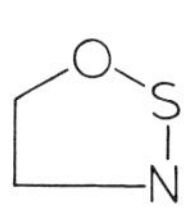	SMG	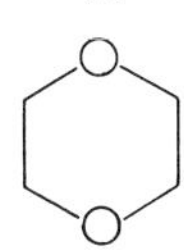
SKC	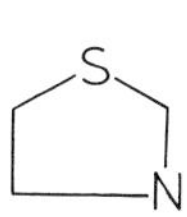	SDX	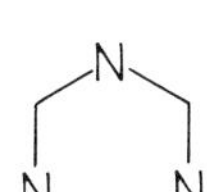

continued

Table 9—continued

GREMAS Term	Typical Structure
SLN	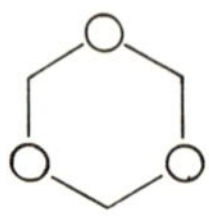
SES	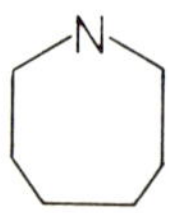
SMK	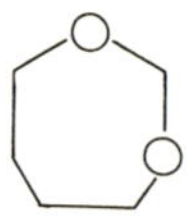

Fused heterocyclics are assigned terms beginning with T rather than S

GREMAS Term	Typical Structure
TDE	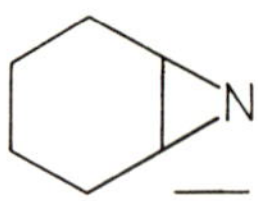
TLB	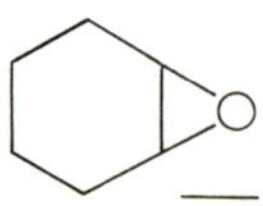
TEA	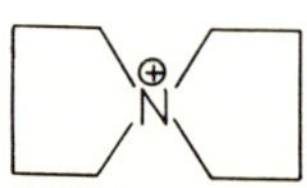
TKC	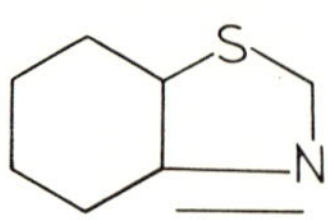

GREMAS Term	Typical Structure
TEL	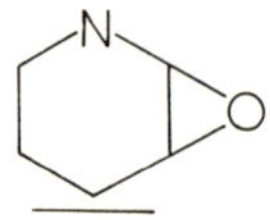
TKP	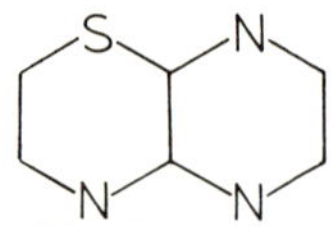
TDX	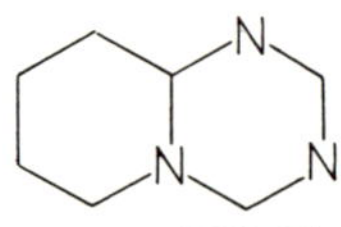

Certain fused heterocyclics are assigned terms beginning with U rather than T

GREMAS Term	Typical Structure
UBH	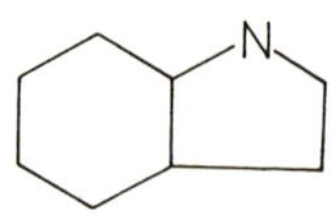
UBI	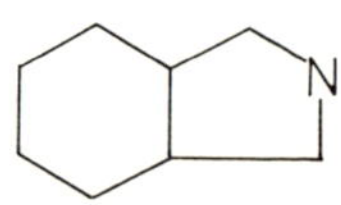
UBE	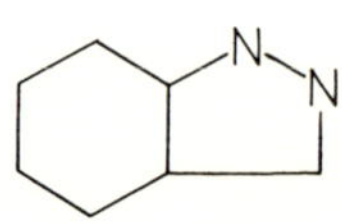
UBD	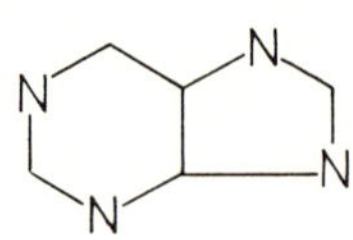

Table 9—concluded

GREMAS Term	Typical Structure	GREMAS Term	Typical Structure
UBF	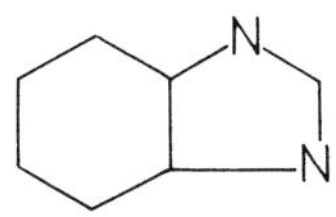	UCG	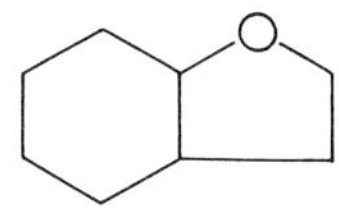
UDA	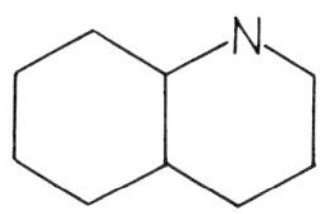	UCH	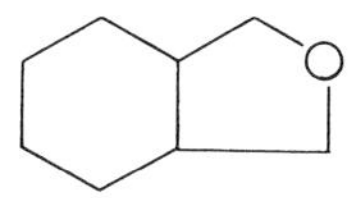
UBA	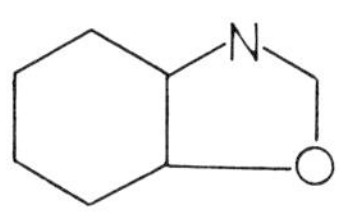	UCI	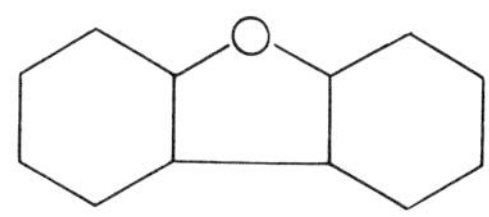
UBL	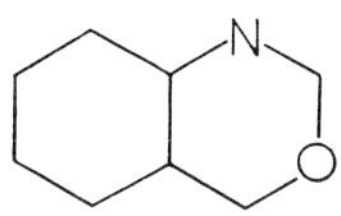	UDE	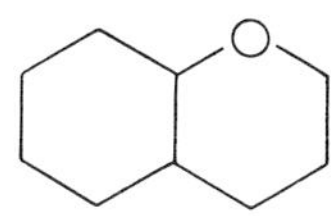
UBN	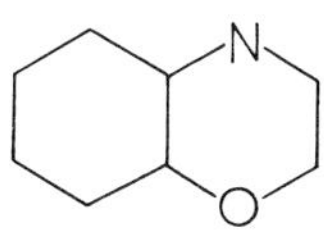	UDF	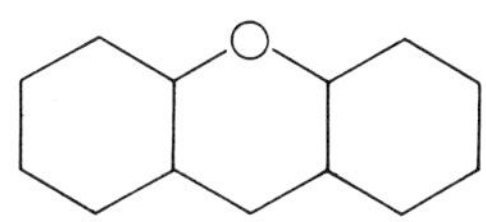
UBC	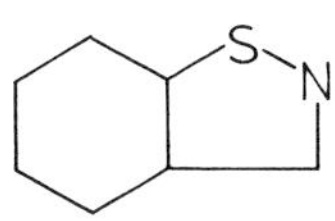	UCK	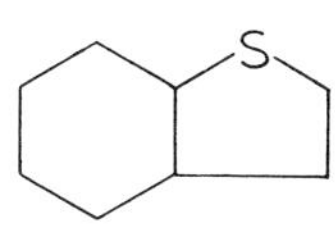
UBB	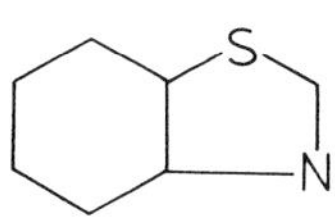	UCL	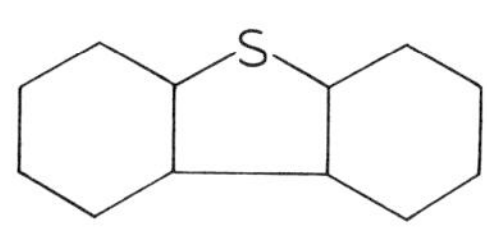
UBO	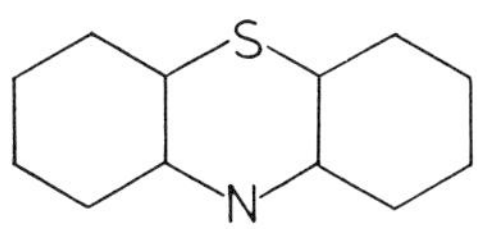	UDL	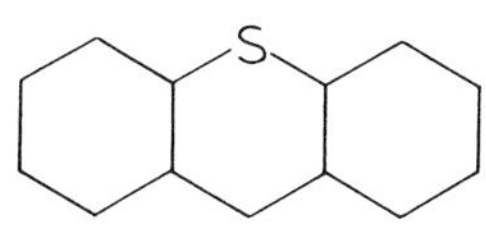

Table 10–GREMAS Terms for Selected Carbocycles

GREMAS Term*	Typical Structure	GREMAS Term*	Typical Structure
SCA	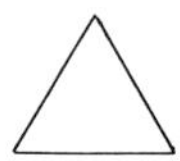	SAD	
SCB		SAF	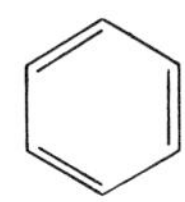
SCC	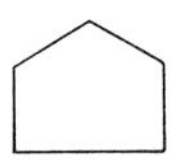	SCG	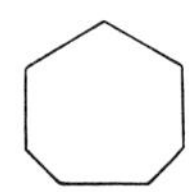
SCE	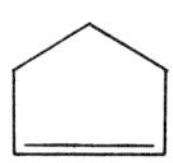	SCK	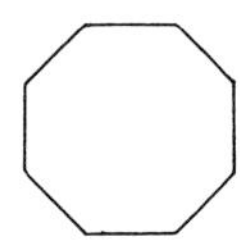
SCF	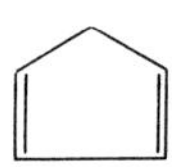	SCN	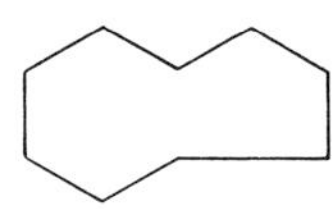
SAA	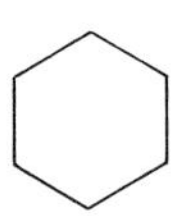	SAK	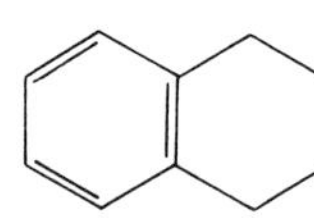
SAB		SAL	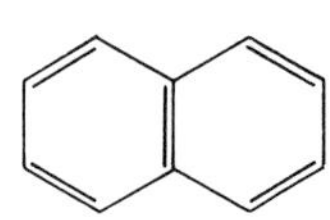
SAC	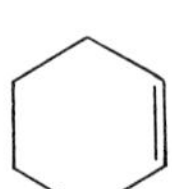	SAN	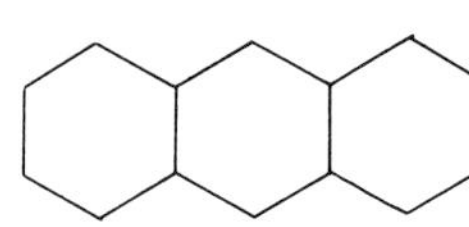

* "S" terms become "T" terms when ring fused to another ring. The degree of unsaturation is specified in the "Y....." terms.

Table 10–continued

GREMAS Term	Typical Structure	GREMAS Term	Typical Structure
SAO	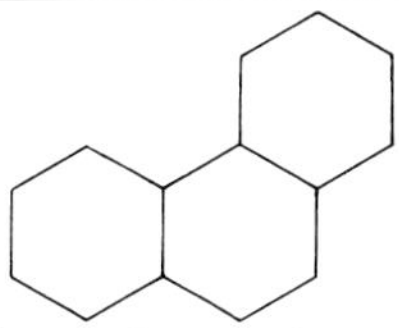	UAC	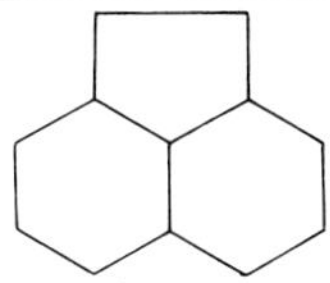
SBB	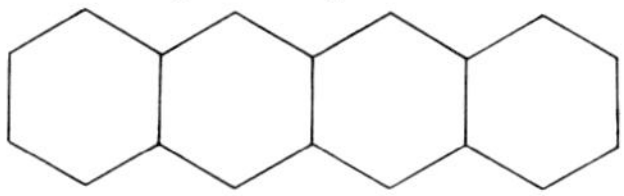	UAD	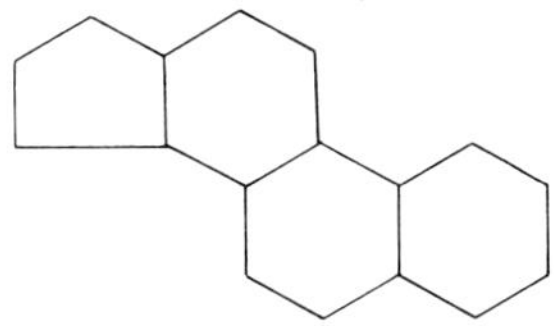
SBE	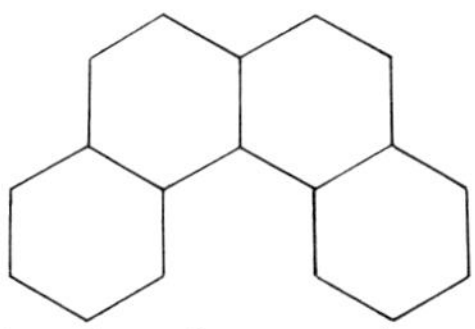	UAE	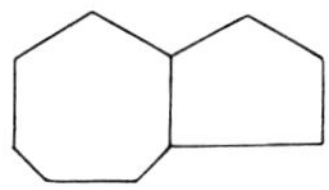
Some fused carbocycles are given "U" terms		UAF*	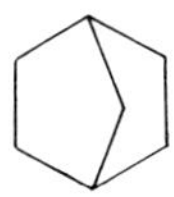
UAA	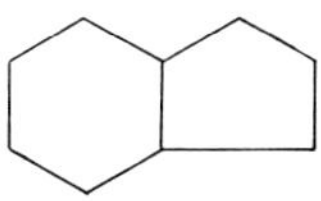	UAG*	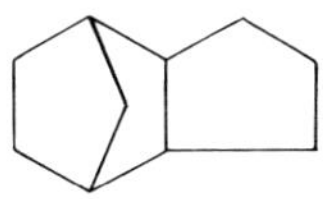
UAB	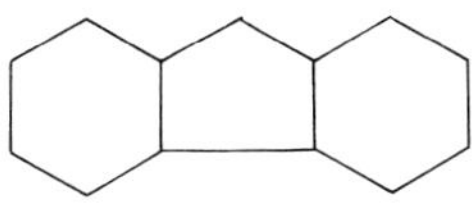		

* The term TRA is also assigned to signify bridge.

Table 11–GREMAS Terms for Certain Features of Ring Systems

GREMAS Term	Structural Feature Indicated
TQA	C-spiro union
TQB	Hetero-spiro union
TRA	Methylene bridge
VAA	1,1 ring fusion (spiro)
VBB	1,2 ring fusion (ortho)
VCB	1,2,3 ring fusion
VDB	1,2,3,4 ring fusion
The following terms apply only to N, O and S heterocycles	
VGA	Carbon-fused rings
VHA	Hetero-fused rings
WGA	Carbon-substituted ring
WHA	Hetero-substituted ring
VGB VHB WGB WHB	As above, but for generic nitrogen heterocycles 
VGE VHE WGE WHE	As above, but for 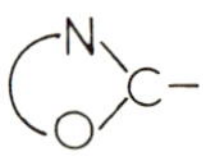

GREMAS Term	Structural Feature Indicated
VGI VHI WGI WHI	As above, but for 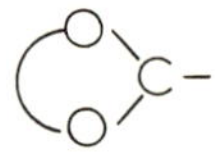
The following terms apply to the particular heterocycles indicated	
SPF	
SPB	
SPC	
SPD	
SPE	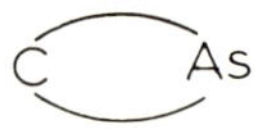
SPG	
SPH	
SPA	
SPK	

Searches using the GREMAS notation are said to be quite economical because the structural fragments required in a typical query are in most cases already preformed in the notation for the filed structures. The precision of GREMAS searches is largely due to the highly selective Y terms and to the fact that both *positive* and *negative* search requirements are permitted. The flexibility in the specification of search requirements is made possible by the particular treatment of Markush structures within the GREMAS system.

SYMBOL	SIGNIFICANCE				
	IUPAC	WISWESSER	HAYWARD	SKOLNIK	GREMAS*
ƀ		Precede multiplicatives, or show multicyclic points in ring systems			
A	A, saturated ring a, for aldehydes (as in EQa)	Generic alkyl group	a, stereo	Methyl group	$>\overset{I}{C}-X$ $X \neq N,O,S,Hal$
B	B, ring system with maximum number of non-cumulative double bonds b, as above, for series of locants	Boron	B, bromine b, stereo	Methenyl group $=CH-$ (chain)	$>\overset{I}{C}-X$ $X = N$
C	C, carbon c, "cis" stereoisomerism	Unbranched carbon with multiple attachments to heteroatoms	Carbon	Methylene group $-CH_2-$ (chain)	$>\overset{I}{C}-X$ $X = NY$ $Y \neq C$
D	*D*, stereo	Proposed: chelate bond and initial symbol of chelate notation		Methenediyl group $=C<$ (chain)	$>\overset{I}{C}-X$ $X = NY$ $Y = N-$
E	Double bond	Bromine	Ethyl (CH_3CH_2-) when not attached to $-CH_2-$	Methene group $=CH_2$ (chain)	$>\overset{I}{C}-X$ $X = O$

*Heteroorientation is denoted by a Roman numeral over the carbon atom in question. Since the GREMAS System is not directly comparable with the other systems cited, we give here the significance of genus only.

Table 12–Summary of Linear Notations Based on Symbology

continued

Table 12—continued

SYMBOL	SIGNIFICANCE				
	IUPAC	WISWESSER	HAYWARD	SKOLNIK	GREMAS
F	Fluorine	Fluorine	Fluorine	Fluorine	$>\overset{I}{C}-X$ X = SY Y = H or C
G		Chlorine		Bromine	$>\overset{I}{C}-X$ X = SY Y = part hetero
H	H, hydrogen h, implied variable hydrogen	When preceded by a locant within ring signs H shows existence of carbon bonded to four other atoms; otherwise H means hydrogen	Hydrogen	Hydrogen	$>\overset{I}{C}-X$ X = halogen
I	Iodine	Iodine	Iodine	Iodine	$>\overset{II}{C}$ ketones and derivatives
J		Generic halogen; terminal symbol of ring description		Methylidyne group at bridgehead or ring fusion point	
K		Nitrogen bonded to more than three other atoms	Chlorine	Carbonyl group	$>\overset{II}{C}$ thioketones and derivs.

SYMBOL	SIGNIFICANCE				
	IUPAC	WISWESSER	HAYWARD	SKOLNIK	GREMAS
L	*L*, stereo	Initial symbol of carbocyclic ring notation	Carbon atom in non-aromatic ring	Chlorine	$\overset{II}{>C}$ aldehydes and derivs.
M	Macrocycle ciphered by assembly notation	>NH	Methyl, not attached to $—CH_2—$	>NH	$\overset{II}{>C}$ thioaldehydes and derivs.
N	Nitrogen	—N—, no hydrogen attachments	N, nitrogen (but see Z) n, heterocyclic nitrogen	—N—, no hydrogen attachments (see Z)	$\overset{III}{—C}$ carboxylic acids and derivs.
O	Unshared oxygen except for —CH=O and >C=O (see Q)	Oxygen, hydrogen free	Oxygen, except V and W		$\overset{III}{—C}$ thiocarboxylic acids and derivs.
P	Phosphorus	Phosphorus	Phosphorus	Phosphorus	$\overset{IV}{C}$ carbonic acid and derivs.
Q	Oxygen shared and in —CH=O and >C=O	—OH (hydroxyl)	—OH (hydroxyl)	—O— =O (except K) (see W)	$\overset{IV}{C}$ thiocarbonic acids and derivs.
R	*R*, stereo r, stereo reference	Benzene ring	Carbon atom in aromatic ring	=C< (fused or bridgehead carbon)	$\overset{o}{C}$ alkanes alkenes alkynes

continued

Table 12 –continued

SYMBOL	SIGNIFICANCE				
	IUPAC	WISWESSER	HAYWARD	SKOLNIK	GREMAS
S	*S*, stereo	Sulfur	Sulfur	Sulfur	Monocyclic rings
T	t, "trans" stereo-isomerism	Initial symbol of heterocyclic ring notation or, within ring signs, like X	Isopropyl, not attached to $-CH_2-$	>C< (fused or bridgehead carbon)	Fused rings
U	Uranium	Double bond (or, when doubled, UU, triple bond)	Uranium	≡C—H	Frequently occurring fused ring systems (entire)
V	Vanadium	Carbonyl connectives (C═O)	Oxo (═O)	≡C–	Points of fusion (rings)
W	Tungsten	Non-linear dioxo group (NO_2, etc.)	Non-linear dioxo group (NO_2, etc.)	Non-linear dioxo group (NO_2, etc.)	Points of substitution (rings) or carbon atom count (chains)
X	X, carboxyl oxygen (═O, OH) x, unknown number	Carbon atom attached to four non-hydrogen atoms		═C═ or >C< (chain)	Bond between hetero-atoms
Y	Triple bond	X–C(H)(X)–X, X═C(X)–X (X not hydrogen or doubly bonded oxygen)	Peripheral ortho-fused carbon atom in a ring	>CH– (chain)	Syntactic term describes sections of molecule

SYMBOL	SIGNIFICANCE				
	IUPAC	WISWESSER	HAYWARD	SKOLNIK	GREMAS
Z	Indicates that hetero atoms follow	$-NH_2$	$-NH_2$	≡N (hydrogen free) =N–	Syntactic term for benzene substitution
0 → 9	Multiplicatives and locants (subscript, multiplicatives; superscript, isotopes and valences)	Multiplicatives when preceded by ƀ (see ƀ). With no leading ƀ, show ring size or length of unbranched aliphatic chains	Ring size when immediately preceding R or L; otherwise multiplicative	Multiplicatives	Used only as species or subspecies, or as multiplicatives in Y and Z terms
:	Used in assembly notation		Non-aromatic double bond in a ring	Atoms between bridgeheads or atoms between spiro atom	
;	Separates groups attached to the same carbon atom		Separates ions		
*		<u>Proposed</u>: symbol for multiplication in inorganic formulas	Marks bridgehead atom (<u>e.g.</u>, 6L*L(1) ...)	Fused or bridgehead atom not carbon	
#				Ionic form	

continued

Table 12—continued

SYMBOL	SIGNIFICANCE				
	IUPAC	WISWESSER	HAYWARD	SKOLNIK	GREMAS
( )	Enclosing marks for compound branches		Enclosing marks for compound branches. Around =, *i.e.*, (=), means "cis"	Encloses substituents	
[ ]	Molecular or ionic entities; common difference in a series of locants		Enclose ionic entities; around =, *i.e.* [=], means "trans"		
{ }					
˥ ˥				Enclose polymer repeating unit	
□				Marks condensed ring structure (point of fusion)	
/	Separates components in assemblies	Terminates action of a multiplicative; encloses repeated symbol groups			
—	Underscores atomic aggregates; marks numbers greater than 9 (*e.g.*, 20)				

SYMBOL	SIGNIFICANCE				
	IUPAC	WISWESSER	HAYWARD	SKOLNIK	GREMAS
&		Terminates branch; b& marks ionic salt, etc.; indicates ring with no H symbol		Followed by atom symbol, indicates those atoms not specially symbolized	
.				Begins and ends syclic structure	Introduces multiplicatives in Y or Z terms
÷			Triple bond		
=			Double bond		
+ −	As superscripts, indicates positive or negative charge				
@			Indicates point of attachment to a new ring		
−	Joins non-consecutive locants	Separator, connector, special uses			Used for padding

Table 12—concluded

It has been shown that the GREMAS notations are readily amenable to the encoding of chemical reactions as well as to structure retrieval applications. The storage and retrieval of chemical reactions data has been an almost entirely neglected area of study in the United States, while the Russians and particularly the Germans[28] have devoted considerable attention to this important aspect of chemical documentation. The work of Lynch in England is also an important contribution to this area.[29]

The preceding table (Table 12) represents a summary comparison of the various notation systems, as far as comparisons are possible. The table is intended to be informative, not evaluative.

NOTES

1. Cf. "Point System," *Webster's New Collegiate Dictionary* (Springfield, Mass.: G. and C. Merriam Co., 1953). See also *Type Faces–An Alphabetical Single-Line Showing* (Easton, Pa.: Mack Printing Co., n.d.).

2. P. N. Craig and H. M. Ebert, "Eleven Years of Structure Retrieval Using the SK&F Fragment Codes," *Journal of Chemical Documentation* 9, no. 3 (1969): 141-146.

3. D. Lefkovitz, "A Chemical Notation and Code for Computer Manipulation," *Journal of Chemical Documentation* 7, no. 4 (1967): 186-191.

4. *Comptes rendus de la 15me Conferénce* (IUPAC, 1949), p. 64.

5. *Chemical Structure Information Handling, A Review of the Literature 1962-1968,* prepared by the Committee on Chemical Information of the Division of Chemistry and Chemical Technology, National Research Council, and published as National Academy of Sciences Pub. No. 1733, Washington, D.C., 1969, pp. 1-8.

6. G. M. Dyson, "A New Notation of Organic Chemistry," Royal Institute of Chemistry Lecture, published jointly with the Chemical Society and the Society of Chemical Industry, 1946.

7. *Comptes rendus de la 14me Conférence* (IUPAC, 1947), p. 64.

8. *Comptes rendus de la 16me Conférence* (IUPAC, 1951), p. 104.

9. *Rules for IUPAC Notation for Organic Compounds,* issued by the Commission on Codification, Ciphering and Punched Card Techniques of the IUPAC, John Wiley & Sons, Inc., New York, 1961.

10. Ibid., p. 23ff.

11. See *Rules for IUPAC Notation for Organic Compounds,* and H. F. Dammers and D. J. Polton, "Use of the IUPAC Notation in Computer Processing

of Information on Chemical Structures," *Journal of Chemical Documentation* 8, no. 3 (1968): 150-160.

12. W. J. Wiswesser, "A Systematic 'Line Formula' Chemical Notation," *ASLIB Proceedings* 5 (1953): 137-147. See also W. J. Wiswesser, *A Line-Formula Chemical Notation* (New York: Thomas Y. Crowell Co., 1954).

13. E. G. Smith, *The Wiswesser Line-Formula Chemical Notation* (New York: McGraw-Hill Book Co., 1968).

14. Ibid., pp. 14-16.

15. Ibid.

16. Ibid., p. 168ff.

17. Ibid.

18. H. W. Hayward, "A New Sequential Enumeration and Line-Formula Notation System for Organic Compounds," *Patent Office Research and Development Reports,* No. 21, U.S. Department of Commerce, Washington, D.C., 1961.

19. H. W. Hayward, H. M. S. Sneed, J. H. Turnipseed, and S. J. Tauber, *Journal of Chemical Documentation* 5, no. 3 (1965): 183-189.

20. H. Skolnik, "A New Linear Notation System Based on Combinations of Carbon and Hydrogen," *Journal of Heterocyclic Chemistry* 6, no. 5 (1969): 689-695.

21. H. Skolnik, "A Correlative Notation System for NMR Data," *Journal of Chemical Documentation* 10, no. 3 (1970): 216-220.

22. A. M. Patterson, L. J. Cappell, and D. S. Walker, *The Ring Index,* 2d ed. (Columbus, Ohio: American Chemical Society, 1960), pp. xv-xvii.

23. Skolnik, "A New Linear Notation System."

24. R. Fugmann, W. Braun, and W. Vaupel, "Zur Dokumentation chemischer Forschungsergebnisse," *Angewandte Chemie* 73, no. 23 (1961): 745-751.

25. R. Fugmann, W. Braun, and W. Vaupel, "GREMAS–ein Weg zur Klassifikation und Dokumentation in der organischen Chemie," *Nachrichten für Dokumentation* 14, no. 4 (1963): 179-190.

26. S. Rössler and A. Kolb, "The GREMAS System, an Integral Part of the IDC System for Chemical Documentation," *Journal of Chemical Documentation* 10, no. 2 (1970): 128-130.

27. E. Meyer, "Versatile Computer Techniques for Searching by Structural Formulas, Partial Structures, and Classes of Compounds," *Angewandte Chemie International Edition in English* 9, no. 4 (1970): 585.

28. R. Fugmann and W. Bitterlich, "Reaktionen Dokumentation mit dem GREMAS System," *Chemiker-Zeitung* 96, no. 6 (1972): 323-329.

29. M. Lynch, J. Harrison, and W. Town, *Computer Handling of Chemical Structure Information* (New York: American Elsevier, 1971), pp. 134-138.

9

Chemical Structure Representation: "Topological Systems"

In preceding chapters various systems belonging to the general class of linear notation systems were described. In this chapter representational systems, which are frequently called "topological" representations of structure, are considered. First, the term "topology" will be analyzed.

TOPOLOGY

Generally, topology may be described as the study of those properties of point sets arising from the notion of connectedness.[1] We shall not attempt mathematical rigor here, but rather an appeal to intuition for an understanding of these concepts. The notion of connectedness can be illustrated in the following way. A circle (Figure 33) is a set of points (all equidistant from a fixed point), which is a connected set because a path exists from every point of the set to every other point of the set (Figure 34). Now, suppose one point were removed from the set: the remaining set of points would still be connected (Figure 35). Finally, consider the removal of two nonadjacent points from the circle: the remaining points are members of two disjoint (disconnected) subsets (Figure 36). In chemistry, aniline is commonly thought of as a connected set (Figure 37), while an aniline complex with TNB

(1,3,5-Trinitrobenzene) might well be considered to consist of two disconnected sets (Figure 38).

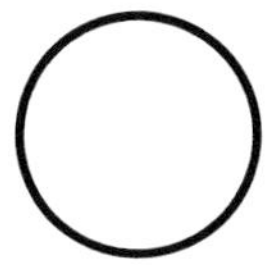

Figure 33–A Connected Set of Points

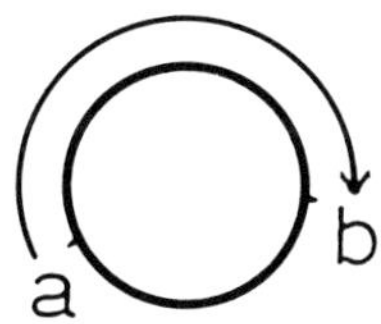

Figure 34–Path Between Any Two Arbitrary Points, a and b, of a Connected Set

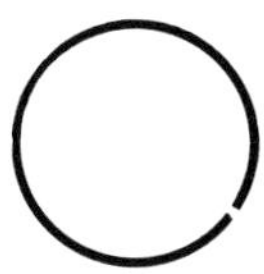

Figure 35–The Connected Set of Figure 33 with One Point Removed

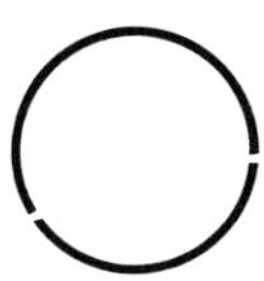

Figure 36–A Disconnected Set of Points (consisting of two connected, disjoint subsets)

Figure 37–A Typical Compound Interpreted as a Connected Set of Nodes (Points)

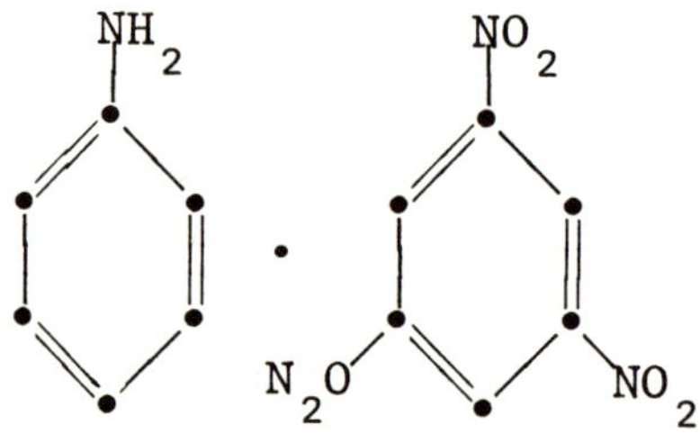

Figure 38–An Aniline/Trinitrobenzene Complex Interpreted as a Disconnected Set (consisting of two connected subsets)

Another important aspect of topology is that of *distance,* which is generally of *no* concern in topology. Imagine, for instance, a triangle drawn on a sheet of rubber (Figure 39). Now imagine that that sheet of rubber is deformed in some way (stretching, bending, etc.). The triangle no longer looks as it did initially (Figure 40), but the new triangle is topologically equivalent to the original one.

The significance of these remarks is that structural representations are almost always topological representations, whether or not the notations are linear. Thus, all of the graphical representations of benzene in Figure 41 are topologically equivalent and some of them can be shown to be equivalent by simple transformations in the plane of the paper. Consequently, we emphasize that all the representations of structure so far considered are topological representations, some representing disconnected sets of nodes of the graphical representation. One

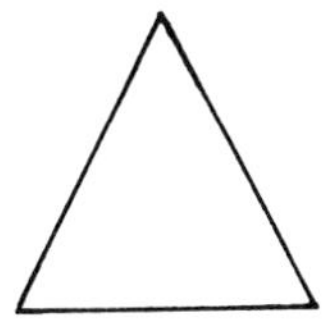

Figure 39—A Triangle Drawn in the Conventional (Euclidean) Manner

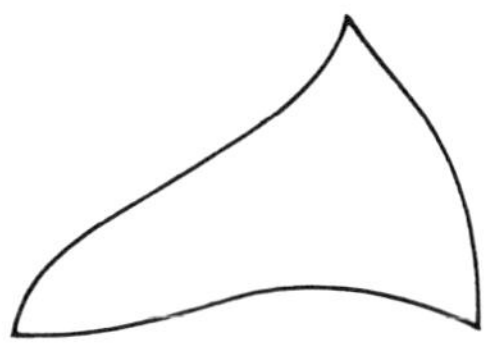

Figure 40—A Figure Topologically Equivalent to the Triangle of Figure 39

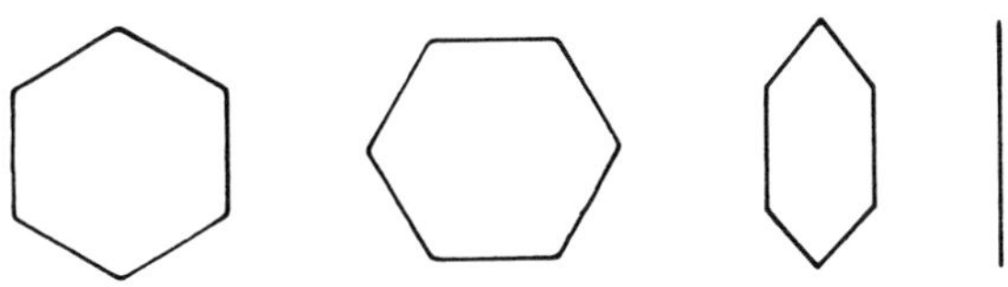

Figure 41—Topologically Equivalent Representations of Benzene

might think of these, respectively, as incomplete and complete representations of structure.

In this chapter, we shall discuss a type of structure representation which has frequently been explicitly labeled "topological," namely that type of representation in which each and every distinguishable feature of a chemical species is *explicitly* identified. These are sometimes called *tabular representations, connectivity tables, matrix representations,* and *incidence matrices.* It has sometimes been said that such representations are especially amenable to computer manipulation, while linear notations are more readily used by humans. It seems more reasonable to characterize "tabular" representations as corresponding to high "specific capability," high "generic capability," and low "compactness" (see Chapter 8). We shall accord them no other distinction.

TERMINOLOGY

For convenience, tabular representations of structures can be referred to as *connection tables.* It is also convenient to speak of chemical structures as graphs. A graph is a set of points called *nodes,* which are connected by lines called *edges.* In chemical structures, the edges are *undirected,* that is, the edges are directionally symmetrical. Undirected graphs may be either *connected* (for every pair of nodes n_i,n_j in the graph, there is a path connecting n_i and n_j) or *disconnected.* Other useful notions are *tree, spanning tree,* and *circuit.* A circuit is a set of edges $\{(V_{i_1},V_{i_2}), (V_{i_2},V_{i_3}), \ldots ,(V_{i_{n-1}},V_{i_n}), (V_{i_n},V_{i_1})\}$ [where an edge is represented by its endpoints (V_{i_j},V_{i_m})], such that if $j \neq m$ then $V_{i_j} \neq V_{i_m}$. A circuit is usually represented in node form, $C_i = (V_{i_1},V_{i_2}, \ldots ,V_{i_n})$. A *tree* is a connected graph which has no circuits. A *spanning tree,* T, is a tree which is a subgraph of a connected graph G, such that every node of G is also a node of T.

HISTORY

The earliest reference to the use of connection-table representations of structure is that by Wheland in his lectures in advanced organic chemistry at the University of Chicago, notes which were copyrighted in 1946[2] and which subsequently formed the substance of a text published in 1949.[3] Wheland showed that acetaldehyde, for example, could be represented as well by the tabular form

	C	C	H	H	H	H	O
C	-	1	1	1	1	0	0
C		-	0	0	0	1	2
H			-	0	0	0	0
H				-	0	0	0
H					-	0	0
H						-	0
O							-

as by the diagrams

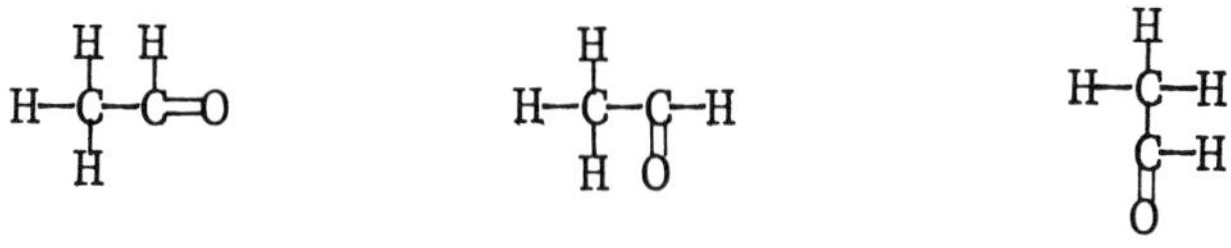

because the diagram had no more geometrical significance than a name or a tabular representation.

In the tabular representations introduced by Wheland, a numerical representation of the type of bond between atoms, whose symbols label the columns and rows of the table, was used (0 = no bond, 1 = single bond, 2 = double bond). The table is clearly symmetrical, so that it would be redundant to fill in the lower (i.e., below the diagonal) portion of the table. Diagonal elements were filled with dashes to signify that no bond was possible "between" an atom and itself.

There is no evidence that this tabular mode of representation was ever developed by Wheland, and it was not until 1951 that it appeared again in the literature. Mooers suggested that a chemical structure could be conveniently represented as a tabulation of atoms and bonds.[4] This tabulation took the form illustrated in Figure 42. Each atom in a structure was assigned a number and each multiple bond was also numbered. A table was then produced which listed the structural features by number and which indicated, again by number, what structural features were adjacent (attached) to the feature under consideration. Finally, the familiar symbols for atoms and bonds were added to the table to complete the representation. Such a representation is, of course, exactly analogous to the matrix representation described by Wheland (though less bulky).

Remarkably, another six years passed before tabular representations again made their appearance in the literature. In 1957, Ray and Kirsch of the National Bureau of Standards were involved in a project to improve literature searching procedures used by the Patent Office.[5] They described the use of Mooers' tabular representations as the basis for a chemical structure search system programmed for operation on the bureau's digital computer, SEAC (N.B.S. Electronic Automatic Computer). Much of the subsequent literature dealing with techniques

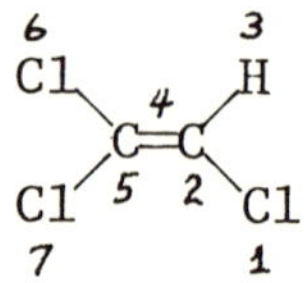

COMPONENT NUMBER	CONNECTION	COMPONENT SYMBOL
1	2	Cl
2	1,3,4	C
3	2	H
4	2,5	=
5	4,6,7	C
6	5	Cl
7	5	Cl

Figure 42–Tabular Representation of a Chemical Structure (after Mooers)

for chemical structure handling incorrectly marks the Ray and Kirsch paper as the origin for tabular representation of structure, even though they credit Mooers with the representation employed in their work.* Nevertheless, it was the paper by Ray and Kirsch which gave impetus to the subsequent developments in tabular representation of structure.

In 1962, Meyer and Wenke described their efforts to refine and extend the tabular representation just discussed.[6] They employed a six-column table; the columns had the following significance:

Column	**Significance**
1	-Atom number (from original structure)
2	-Atom type (numerical code for atom symbol)
3-6	-Numbers of those atoms connected to the atom whose number appears in column 1 of a given row.

*The chronology presented here ignores the literature of mathematics in which tabular representations of graphs may have much earlier precedence.

STRUCTURE

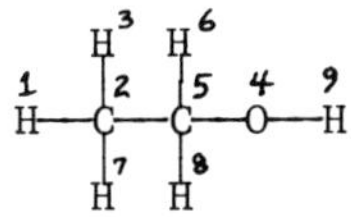

TABLE

ATOM	TYPE	CONNECTIONS			
1	4	2	0	0	0
2	1	1	3	5	7
3	4	2	0	0	0
4	2	5	9	0	0
5	1	8	2	6	4
6	4	5	0	0	0
7	4	2	0	0	0
8	4	5	0	0	0
9	4	4	0	0	0

ATOM-TYPE CODES C=1
O=2
H=4

Figure 43—Connection Table Developed by Meyer and Wenke

This method of representation is illustrated in Figure 43. One important facet of this representation is that no bonds are cited in the table. Single bonds are always assumed; multiple bonding and special features of the structures, such as ring membership, charge, and isotopic mass, are included in separate subtables. These subtables are identified by a numerical "tag" which indicates the type of feature possessed by the atoms whose numbers appear in the list (see Figure 44). As will be seen, the tables developed by Meyer and Wenke bear considerable resemblance to the form of table developed subsequently by Gluck.[7]

The work of Meyer and his co-workers at B.A.S.F. and of his colleagues Fugmann and Möhring of Hoechst and Baeyer, respectively, has led to many significant accomplishments, including the development of an operational optical scanner for the input of hand-drawn structures;[8] to the computer generation of linear notations (fragment codes) from connection tables; and to an integrated structure and text retrieval system. The development of what has become an international

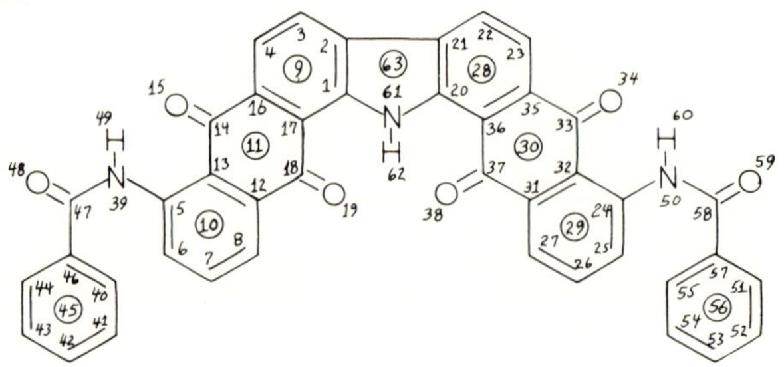

ATOM	TYPE	CONNECTIONS			
1	1	2	17	61	0
2	1	1	3	21	0
3	1	2	4	0	0
4	1	3	16	0	0
5	1	6	13	39	0
6	1	5	7	0	0
7	1	6	8	0	0
8	1	7	12	0	0
9	0	0	0	0	0
10	0	0	0	0	0
11	10	0	0	0	0
12	1	8	13	18	0
13	1	5	12	14	0
14	1	13	15	16	0
15	2	14	0	0	0
16	1	4	14	17	0
17	1	1	16	18	0
18	1	12	17	19	0
19	2	18	0	0	0
20	1	21	36	61	0
21	1	2	20	22	0
22	1	21	23	0	0
23	1	22	35	0	0
24	1	25	32	50	0
25	1	24	26	0	0
26	1	25	27	0	0
27	1	26	31	0	0
28	0	0	0	0	0
29	0	0	0	0	0
30	10	0	0	0	0
31	1	27	32	37	0
32	1	24	31	33	0
33	1	32	34	35	0
34	2	33	0	0	0
35	1	23	33	36	0
36	1	20	35	37	0
37	1	31	36	38	0

ATOM	TYPE	CONNECTIONS			
38	2	37	0	0	0
39	3	5	47	49	0
40	1	41	46	0	0
41	1	40	42	0	0
42	1	41	43	0	0
43	1	42	44	0	0
44	1	43	46	0	0
45	0	0	0	0	0
46	1	40	44	0	0
47	1	39	46	0	0
48	2	47	0	0	0
49	4	39	0	0	0
50	3	24	58	60	0
51	1	52	57	0	0
52	1	51	53	0	0
53	1	52	54	0	0
54	1	53	55	0	0
55	1	54	57	0	0
56	0	0	0	0	0
57	1	51	55	58	0
58	1	50	57	59	0
59	2	58	0	0	0
60	4	50	0	0	0
61	3	1	20	62	0
62	4	61	0	0	0
63	10	0	0	0	0

Double bond pairs

12	50	14	15	18	19	33	34		
		37	38	47	48	58	59		

Ring membership

9	70	9	1	17	16	4	3	2
		10	5	13	12	8	7	6
		11	12	13	14	16	17	18
		28	20	21	22	23	35	36
		29	24	25	26	27	31	32
		30	31	37	36	35	33	32
		45	40	41	42	43	44	46
		56	51	52	53	54	55	57
		63	1	2	21	20	61	0

TYPE CODES

1 = C
2 = O
3 = N
4 = H
0 = aromatic ring
10 = non-aromatic ring
50 = double bond
70 = enumeration of ring members (4, 5 & 6 membered)

Figure 44—A Meyer-Wenke Connection Table

This table includes special features (rings and double bonds); hydrogen attached to carbon has been ignored.

chemical information system can also be credited to the impetus of these groups.[9]

At about the same time that the work of Meyer and Wenke was under way, Spialter reported a "new unambiguous universal symbolism" for chemical compounds, which he called the Atom Connectivity Matrix (ACM).[10] The ACM is exactly analogous to Wheland's matrix, except that the atom symbols are placed along the principal diagonal, rather than at the heads of columns and rows (see Figure 45). The off-diagonal elements of the ACM contained bond symbols or other features of structure (transition states, etc.) as desired. One novel* feature introduced by Spialter was the characteristic polynomial of the ACM (called the ACMCP), obtained by evaluating the ACM as a determinant.[13] Thus, hydrogen cyanide would have the ACM and ACMCP shown below. 3-Methylcyclopropene serves to further exem-

Structure **ACM** **ACMCP**

H–C≡N

$$\begin{vmatrix} C & 1 & 3 \\ 1 & H & 0 \\ 3 & 0 & N \end{vmatrix} = HCN-9H-N$$

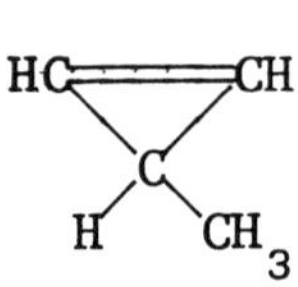

$$\begin{vmatrix} C & 1 & 0 & 0 \\ 1 & C & 1 & 1 \\ 0 & 1 & C & 2 \\ 0 & 1 & 2 & C \end{vmatrix} = C^4-7C^2+4C+4$$

Figure 45–Illustrations of Spialter's ACM and ACMCP

*Imputation of novelty is a hazardous business. There is a suggestion in Vleduts and Seifer[11] that the concept of a characteristic polynomial for chemical structures was described as early as 1936 by a Russian mathematician, N. S. Kurnakov.[12]

plify the concept of the ACMCP. While the ACMCP is a unique (but see Balaban and Harary[14]) representation of structure, the structure cannot be regenerated from the polynomial, so that the ACMCP serves only as a "registry number," or identifier for a chemical substance.

Another work of note in this brief history is that of Dyson, Cossum, Lynch, and Morgan, published in 1963.[15] The aspect of their research which is of greatest interest here is the development of procedures for the interconversion of linear notations *via* a matrix representation of the structure of a substance and for the derivation of molecular formulas from linear notations, particularly IUPAC linear notations (see Chapter 8).[16] The research was also concerned with substructure searching (the identification and retrieval of those structures, from a file of structures, which possess some common structural characteristic[s]), and the matrix form of representation again played an important role. An illustration should clarify the tabular representations employed by these workers and how they are derived from IUPAC notations. Consider, for instance, the following example:

Structure

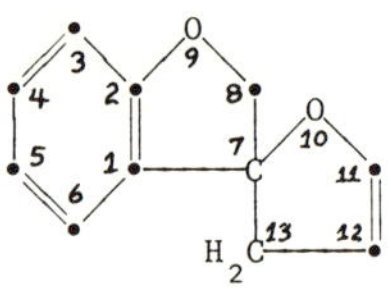

Name

Spiro[benzofuran-3(2*H*),2'(5'*H*)-furan]

IUPAC Notation

B65$_2$17-7ZQ9<u>10</u>h<u>13</u>

Table

	RANK	RING	RING SIZE	MATRIX†		MOL FORM		
Cyclic Portion	I	B	6	1 2 3 4 5 6		C	H	O
	II	B	5	1 $\hat{7}$ 8 $\hat{9}$	2			
	III	B	5	7 $\hat{10}$ 11 12 $\hat{13}$	7	13		
							14	
		SUBSTITUENT	LOCANT					
Acyclic Portion	IV	ZQ	9			−1	−2	1
	V	ZQ	10			−1	−2	1
	VI	h	13					
				Molecular Formula		C_{11}	H_{10}	O_2

†Fusion locants appear outside the matrix at the row corresponding to the ring to which they belong.

The table is formed by scanning the IUPAC linear notation in a left-to-right manner, entering as a rank in the table each distinct feature of the notation (expanding repeated elements). The example compound above contains three rings, all of which are of B type (see Chapter 8). The six-ring is first encountered in the notation and is entered as rank I in the table. The matrix row 1 is formed by listing all of the atoms of this first ring. Rank II is formed for the first five-membered ring, but only those atoms of the ring which do not already appear in the matrix are entered, the remainder being placed just outside the matrix. Rank III is formed similarly. The molecular formula is now partially developed by treating the entire structure as a hydrocarbon. In this example, the highest locant (13) in the matrix is entered into the "carbon" column of the molform subtable. Hydrogen is computed using the formula

$$H = 2m - (2n_1 + n_2 - 2)$$

where H is the hydrogen number, m is the number of carbons in the "carbon" column, n_1 is the number of rows in the matrix, n_2 is the number of locants within the matrix not marked (˷) nor isolated (locants between marked locants, either within or without the matrix). For our example,

$$\begin{aligned} H &= 26 - (6 + 5 - 2) \\ &= 14 \end{aligned}$$

Marks are adduced to any locants in the matrix which refer to h or H terms in the IUPAC notation (see Chapter 8); to any locants which occur v-2 times (v = valency of atom indicated by the locant) in the matrix or substituent-locant list; and to any locants of heteroatoms which occur v-1 times anywhere in the locant series (E accounts for two of these occurrences). The nonring terms of the notation are also set out as ranks of the table and contribute (as shown in the example) to the development of the molecular formula.

Dyson, Cossum, Lynch, and Morgan recognized the merit of their tabular representation of structure as an interface language through which communication between various systems could be effected. This idea has not been exploited to any significant degree. Some translating systems have been developed (see, for example, Garfield's work), but a generalized interface language has not been employed. The work of these authors was never fully developed or implemented; instead it was supplanted by the developments described by Gluck and by Morgan.

During the early sixties, Gluck and his colleagues at DuPont undertook the development of a tabular representation of structure and of a computer system through which a unique and unambiguous connection table convenient for storage and retrieval purposes could be derived.

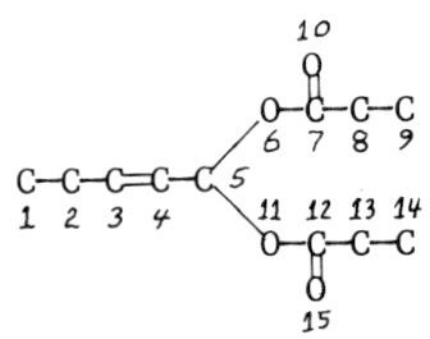

Atom No.	Atom Code	Group	Bond 1 Type	Bond 1 Atom No.	Bond 2 Type	Bond 2 Atom No.	Bond 3 Type	Bond 3 Atom No.	Bond 4 Type	Bond 4 Atom No.	Bond 5 Type	Bond 5 Atom No.
1	C		1	2								
2	C		1	3								
3	C		2	4								
4	C		1	5								
5	C		1	6	1	11						
6	O		1	7								
7	C		1	8	2	10						
8	C		1	9								
9	C											
10	O											
11	O		1	12								
12	C		1	13	2	15						
13	C		1	14								
14	C											
15	O											

Figure 46–Input Connection Table of the Type Used by Gluck

Input to the computer system was prepared clerically from hand-drawn structures, using a form of the type shown in Figure 46. The form was obviously devised to facilitate its conversion to punched-card form for computer input. Once the structure had been input to the computer system, a number of error checks (editing procedures) were carried out to insure the accuracy of the input. Once the input data had passed the edit checks, a canonical* connection table was generated. It will be important to our subsequent discussion to understand how the canonical form is produced, but for now we will be satisfied with a general description. The connection table (see Figure 47, for example) was reordered in such a way that the atoms of highest degree (connectivity), then with the lexicographically earliest atom symbol and, finally, with the earlier bond symbol, came earliest in the table. There would frequently result ties (two atoms vying for the same position in the table) which had to be resolved. The atoms of the structure were numbered to reflect their new positions in the table. An iterative procedure was then employed to break all ties.** Once the canonical table had been generated, it was converted into a more compact form, which Gluck called a "compact list."[17] The compact list was formed by once again renumbering the structure according to an algorithm given in full in the next section. Figure 48 gives the canonical connection table and compact list for the diester shown.

The details of the work carried out at DuPont were made available to the American Chemical Society (ACS). The basic techniques, following modification and refinement, were adopted by the ACS through its information processing division, the Chemical Abstracts Service (CAS), as the basis for a full-scale chemical information system. The details of the CAS effort are dealt with, in a somewhat tutorial fashion, in the next section.

*From *canon law,* a critical standard or criterion. To our knowledge, this work represents the first attempt to generate a unique connection table from unambiguous but nonunique input.

**It can be shown that this tie-breaking procedure may in certain cases not yield the desired canonical form of the table and the subsequent input of the same compound *could* result in a second (different) table being entered into the files. This fault was obviated by Morgan in later work (see next section).

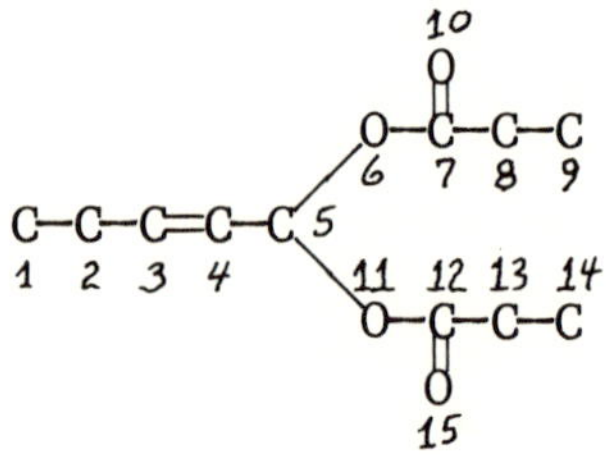

		Bond 1		Bond 2		Bond 3	
Atom	Code	Type	No.	Type	No.	Type	No.
1	C	1	2				
2	C	1	3	1	1		
3	C	2	4	1	2		
4	C	1	5	2	3		
5	C	1	6	1	11	1	4
6	O	1	7	1	5		
7	C	1	8	2	10	1	6
8	C	1	9	1	7		
9	C	1	8				
10	O	2	7				
11	O	1	12	1	5		
12	C	1	13	2	15	1	11
13	C	1	14	1	12		
14	C	1	13				
15	O	2	14				

Figure 47–Non-unique Connection Table Developed by Gluck's System

This brief history traces those developments which we believe led most directly to the connection table representation described by Morgan in 1965.[18] No doubt, many significant works have been left out, but a more detailed history is not possible here. In addition, this chapter largely ignores any work relating to connection table representation of structure published after Morgan's paper. We felt safe in doing this because very little work has been directed toward connection

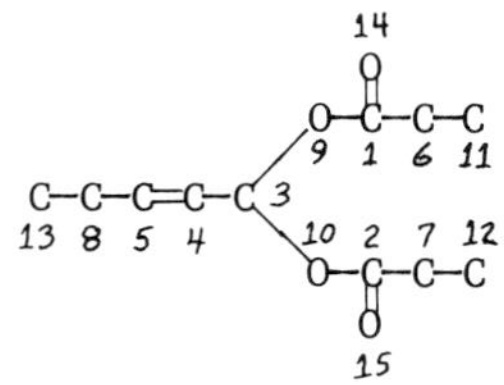

CANONICAL FORM

		Bond 1		Bond 2		Bond 3	
Atom	Code	Type	No.	Type	No.	Type	No.
1	C	2	14	1	6	1	9
2	C	2	15	1	7	1	10
3	C	1	4	1	9	1	10
4	C	2	5	1	3		
5	C	2	4	1	8		
6	C	1	1	1	11		
7	C	1	2	1	12		
8	C	1	5	1	13		
9	O	1	1	1	3		
10	O	1	2	1	3		
11	C	1	6				
12	C	1	7				
13	C	1	8				
14	O	2	1				
15	O	2	2				

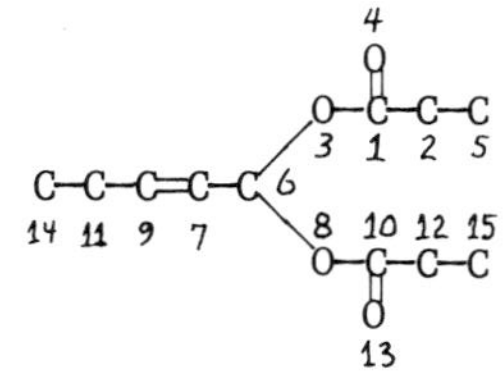

COMPACT LIST

Atom	Code	Attached Atom	Bond Type
1	C	2	-
2	C	5	1
3	O	6	1
4	O	-	2
5	C	-	1
6	C	7	1
7	C	9	1
8	O	10	1
9	C	11	2
10	C	12	1
11	C	14	1
12	C	15	1
13	O	-	2
14	C	-	1
15	C	-	1

Figure 48–Canonical Connection Table and Compact List as Produced by Gluck's Procedures. In the compact list, "bond type" refers to the bond joining the atom having this row number to the lowest numbered atom attached thereto.

tables per se. Most work has been directed toward applications of various forms of structure representations; such work is discussed elsewhere. Perhaps the most significant work on structure representation subsequent to 1965 was that concerned with configurational and conformational aspects of structure (no longer topological but geometric).[19] This is beyond the scope of this text.

THE CONNECTION-TABLE REPRESENTATION EMPLOYED BY THE CHEMICAL ABSTRACTS SERVICE

The work of Gluck and of Morgan led the Chemical Abstracts Service and the DuPont Company to develop and implement a system for the registration of chemical compounds. The system developed by CAS became known as the Chemical Compound Registry System (herein called the Registry System), and that developed by DuPont became known as the Chemical Structure Storage and Search System (CS^4). Here we shall discuss only the Registry System, but much of this discussion also applies to the CS^4.

The purpose of the Registry System was to enable the unique and unambiguous identification of a chemical substance and to provide a representation suitable for storage and subsequent retrieval, all by computer. Several years of independent research by DuPont and by CAS ultimately led both organizations to adopt a connection-table representation of structure as the basis for achieving these goals.*

In order to achieve unique and unambiguous identification (i.e., representation) of chemical structures, it was necessary to develop some rules, just as for linear notation systems, by means of which a unique and unambiguous connection table could be generated. Let us consider what is involved in developing such rules. First, the question of ambiguity must be resolved. Since atom symbols may occur repetitively within a chemical structure, it is necessary to give each atom a unique identity.

*There is a good deal of history pertinent to this statement, including the role of the Committee on Modern Methods of Handling Chemical Information (established by the National Research Council) and the working relationships between DuPont and CAS.[20]

This is most easily accomplished by assigning to each atom in a structure a number different from any other number assigned within that structure. It is also convenient to assign consecutive numbers and to start with the number 1. Wheland satisfied this requirement by placing an atom symbol in a particular row or column of a table, viz.:

	C	C	H	H	H	H	O
C	-	1	1	1	0	0	1
C		-	0	0	1	1	1
H			-	0	0	0	0
H				-	0	0	0
H					-	0	0
H						-	0
O							-

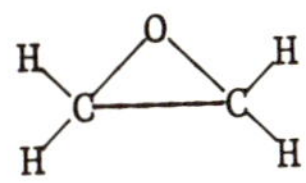

Hence, the first C is interpreted as C_1, the second as C_2, the first H as H_3, and so on. In this way, we note, for example, that H_6 is connected to C_2, and H_5 is connected to C_1, rather than, simply, C is connected to H. Thus, we require a rule which associates with each atom of a molecule a unique number, one of a consecutive sequence of numbers, 1, 2, 3, ...

We must also require that each type of bond be unambiguously identified. This can be accomplished by associating a number or other symbol with each bond type. The combination of atoms and bonds denoted in this way will serve unambiguously to define the topology of any molecular species of known structure. Consequently, Wheland's tables are unambiguous representations of structure. But it can easily be seen that our rules for labeling atoms and bonds (nodes and edges) are not sufficient for uniqueness of a connection table. For instance, the atoms of acetylene, **H—C≡C—H** can be assigned numbers in any of twenty-four ways:

1	2	3	4	2	1	3	4	3	1	2	4	4	1	2	3
1	2	4	3	2	1	4	3	3	1	4	2	4	1	3	2
1	3	2	4	2	3	1	4	3	2	1	4	4	2	1	3
1	3	4	2	2	3	4	1	3	2	4	1	4	2	3	1
1	4	2	3	2	4	1	3	3	4	1	2	4	3	1	2
1	4	3	2	2	4	3	1	3	4	2	1	4	3	2	1

Hence, a rule is required for choosing just one of the ways. In fact, it is easy to show that for a molecule of *n* distinct atoms (i.e., a graph of *n* distinct nodes) there are actually n! ways of numbering or labeling the atoms (nodes). In the example above (acetylene, n = 4), there are four candidates for the number 1. For each choice for the node numbered 1, there remain three candidates for number 2. For each choice for the node numbered 2, there remain two candidates for number 3 and, finally, for each node numbered 3, there remains one candidate for number 4. Hence, the total number of labelings is 4 x 3 x 2 x 1 = n ! = 24, corresponding to twenty-four topologically equivalent connection tables. To choose from among these a preferred (unique) connection table, the following rule may be used. Upon producing the n! connection tables for a structure, sort the tables in ascending lexicographic order and choose as preferred that one which comes first in the sorted set of tables (essentially a method suggested by Mooers[21]). The application of this rule to acetylene (to which the two previously stated rules have already been applied) is illustrated in Table 13. The reader can easily satisfy himself that whenever the three rules which we have stated are applied to a chemical structure, the same preferred connection table will always be obtained.

It takes little imagination to see that the third rule would, for even the simplest of compounds, require enormous amounts of time for its implementation, even if the processes were carried out by computer. Table 14, which contains values of n!, demonstrates the dramatic increase in the number of topologically equivalent connection tables obtainable for molecules of small to moderate size (in terms of numbers of atoms). We must therefore seek a more efficient, but equally effective, method for deriving a unique connection table.

One way of overcoming the practical difficulty of rule 3 was intro-

Table 13–The 24 Topologically Equivalent Connection Matrices for Acetylene Sorted in descending order (across, then down). Here C precedes H, C precedes Ċ, and so on. The dot is used to facilitate visual comparison of the matrices.

	C	Ċ	H	Ḣ
C	-	3	1	0
•C		-	0	1
•H			-	0
H				-

	C	Ċ	Ḣ	H
C	-	3	0	1
•C		-	1	0
•H			-	0
H				-

	C	H	Ċ	Ḣ
C	-	1	3	0
H		-	0	0
•C			-	1
•H				-

	C	Ḣ	Ċ	H
C	-	0	3	1
•H		-	1	0
•C			-	0
H				-

	C	H	Ḣ	Ċ
C	-	1	0	3
H		-	0	0
•H			-	1
•C				-

	C	Ḣ	H	Ċ
C	-	0	1	3
•H		-	0	1
H			-	0
•C				-

	Ċ	C	H	Ḣ
•C	-	3	0	1
C		-	1	0
H			-	0
•H				-

	Ċ	C	Ḣ	H
•C	-	3	1	0
C		-	0	1
•H			-	0
H				-

	Ċ	H	C	Ḣ
•C	-	0	3	1
H		-	1	0
C			-	0
•H				-

	Ċ	Ḣ	C	H
•C	-	1	3	0
•H		-	0	0
C			-	1
H				-

	Ċ	H	Ḣ	C
•C	-	0	1	3
H		-	0	1
•H			-	0
C				-

	Ċ	Ḣ	H	C
•C	-	1	0	3
•H		-	0	0
H			-	1
C				-

	H	C	Ċ	Ḣ
H	-	1	0	0
C		-	3	0
•C			-	1
•H				-

	H	Ċ	C	Ḣ
H	-	0	1	0
•C		-	3	1
C			-	0
•H				-

	H	C	Ḣ	Ċ
H	-	1	0	0
C		-	0	3
•H			-	1
•C				-

	H	Ċ	Ḣ	C
H	-	0	0	1
•C		-	1	3
•H			-	0
C				-

	H	Ḣ	C	Ċ
H	-	0	1	0
•H		-	0	1
C			-	3
•C				-

	H	Ḣ	Ċ	C
H	-	0	0	1
•H		-	1	0
•C			-	3
C				-

	Ḣ	C	Ċ	H
•H	-	0	1	0
C		-	3	1
•C			-	0
H				-

	Ḣ	Ċ	C	H
•H	-	1	0	0
•C		-	3	0
C			-	1
H				-

	Ḣ	C	H	Ċ
•H	-	0	0	1
C		-	1	3
H			-	0
•C				-

	Ḣ	Ċ	H	C
•H	-	1	0	0
•C		-	0	3
H			-	1
C				-

	Ḣ	H	C	Ċ
•H	-	0	0	1
H		-	1	0
C			-	3
•C				-

	Ḣ	H	Ċ	C
•H	-	0	1	0
H		-	0	1
•C			-	3
C				-

Table 14–Selected Values of *n* and Their Factorials (*n!*)

n	$n!$
1	1
2	2
3	6
4	24
5	120
6	720
7	5040
8	40320
9	362880
10	3628800
⋮	⋮
15	1307674368000
⋮	⋮
20	2432902008176640000

duced by Gluck, although he did not employ it for our purpose.* The following algorithm provides a method for labeling (numbering) the nodes of a graph which substantially reduces the number of connection tables that would have to be generated in order to identify a preferred table.

(1) Consider a graph G consisting of a set N of nodes. The number of nodes in N is $|N|$.

(2) Choose any node $v \in N$ and label it node number 1.

(3) Set $q = 1, m = 2$.

(4) Now consider the set of nodes adjacent (attached) to node number q, which have not been previously labeled. Assign to

*Gluck's numbering algorithm is introduced here at an earlier stage in the canonization process, and has been slightly modified to allow for choices which he did not have to face.

each node of this set the numbers m, m + 1, m + 2, . . . , t, in any order, as far as necessary to label all nodes in the set.

(5) Set m = t + 1, q = q + 1. If q > |N|, terminate the procedure. Otherwise return to step 4.

The application of this algorithm to a simple structure is illustrated in Figure 49.

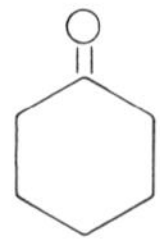

ITER-ATION	STEP				
	1	2	3	4	5
1	N = 7	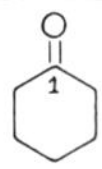	q = 1 m = 2	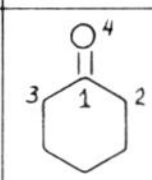	m = 5 q = 2
2				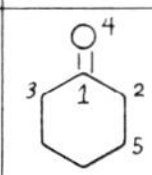	m = 6 q = 3
3				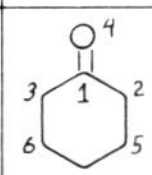	m = 7 q = 4
4				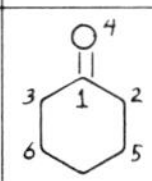	m = 7 q = 5
5				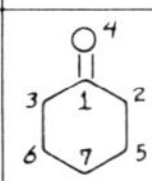	m = 8 q = 6
6					m = 8 q = 7
7					m = 8 q = 8 stop

Figure 49–Illustration of the Application of the Node-labeling Algorithm to Cyclohexane

One may ask: "How many fewer tables must be generated when this technique of numbering is employed?" A simple answer cannot be given, but one or two examples should serve to indicate the magnitude of the reduction which the method affords. For acetylene, the number of tables to be generated is reduced from 24 to 6. For cyclohexane, ignoring hydrogen, n = 6 and n ! = 720. The above algorithm reduces to 12 the number of tables to be generated. Thus, this numbering algorithm affords a great reduction in the number of connection tables which must be produced in order to find a preferred table. Can we do better still? The answer is "yes," since our numbering algorithm presently has no basis for choosing number 1, nor any basis for choice at any other level when more than one node (atom) exists at the level. Morgan recognized this and also realized that, because all the atoms of a structure did not have the same number of connections to other atoms in the structure (that is, the nodes of the graph were not all of like degree), connectivity might be usefully employed in deciding to which atom to assign a number whenever a choice exists. But the technique finally developed by Morgan also took advantage of the "connection environment" of the atoms, as will be seen by inspecting the following algorithm.

(1) Consider a graph G of $|N|$ nodes.

(2) Assign to each node $v_i \in N$ $(1 \leqslant i \leqslant |N|)$ a connection number q_i. Set $q_i = 1$ initially for all values of i. Set k = 1 (k represents the number of iterations through the procedure).

(3) Form a set K_k of unique connection values, q.

(4) If k = 1, go to step 5. Otherwise, if $|K_k| \leqslant |K_{k-1}|$, terminate the procedure and let the connectivities of the nodes of G be their connection numbers computed in iteration k-1.

(5) Otherwise, set k = k + 1. Then, for each $v_i \in N$ with connection number q_i, set $q_i = \sum_{\forall a} q_a$, where q_a is the connection number computed in iteration k-1 of node v_a *adjacent* to node v_i. Add the connection number q_i to the set K_k if it is not already contained in K_k.

(6) When the q's for all v ϵ N have been computed in this way, return to step 4.

The application of this algorithm is depicted in Figure 50. With the connectivities of the nodes of a graph available, the numbering algorithm given earlier may be applied in a more deliberate way, as follows. When a choice exists in the assignment of a node number, select that node which has the highest connectivity. If the connectivity cannot be

k	STRUCTURE WITH COMPUTED CONNECTIVITIES, q	UNIQUE VALUES OF q	$\lvert K_k \rvert$
1	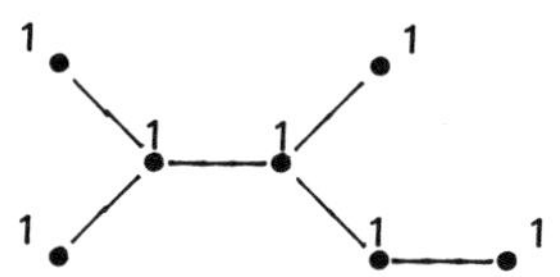	1	1
2	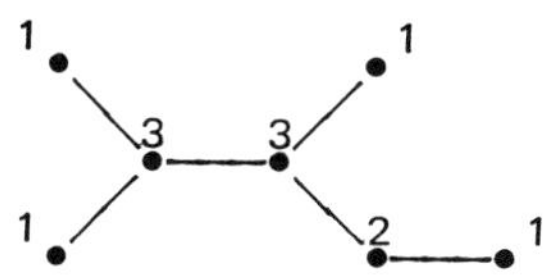	1,2,3	3
3	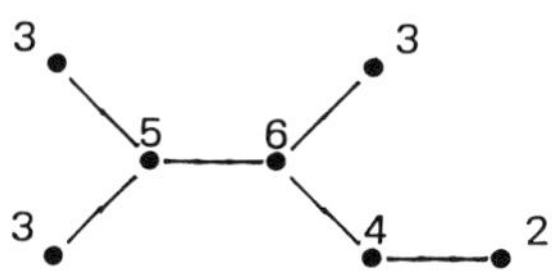	2,3,4,5 6	5
4	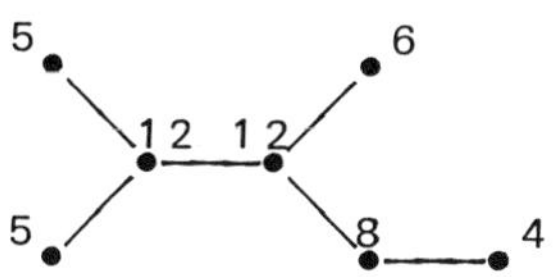	4,5,6,8 12	5

Figure 50—Illustration of the Application of the Algorithm for Computing Connection Values for the Nodes of a Graph

used to make a choice (i.e., there exist two or more atoms with the same connectivity which could receive the number), generate all the alternate numberings, as before. Again, using acetylene as an example, we see that atom connectivities are

$H{-}C{\equiv}C{-}H$

CONNECTIVITY	→	1 2 2 1
		$H{-}C{\equiv}C{-}H$
NODE NUMBER	→	3 1 2 4
(or)	→	4 2 1 3

and that exactly two numberings of the nodes are possible using the connectivity of a node as a selection criterion.

In general, it will not be possible to say what degree of reduction this latter algorithm affords because of its dependence upon the variety of node connectivities extant in a graph. For instance, no reduction at all is achieved for cyclohexane, whereas for 2,2-dimethylbutane, the number of tables to be produced is reduced from 46 (using only the numbering algorithm) to 6. Thus, we now have a method for selecting from the set of possible connection tables for a structure exactly one preferred table, and the method requires the actual generation of only a very few of the tables of this set. Furthermore, since the selection criteria used depend upon invariant properties of a graph (the way the nodes are interconnected and the number of connections to each node), we may be sure that a table selected as preferred at one point in time will also be selected as preferred at any other point in time, when the algorithms described are applied to the same structure. It is also clear that different structures will always be represented by different (distinct) connection tables. Hence, the method produces unique and unambiguous representations of chemical structures.

So far we have depended upon the tables devised by Wheland for our illustrations. Thus, through the application of the rules and algorithms presented above, we find that, for acetylene, tables 1 and 8 of Table 13 are of equal preference and are in fact isomorphic.* Hence, either may

*Two structures (strictly, groups) may be said to be *isomorphic* when there is a one-to-one correspondence between their nodes which preserves the order

be arbitrarily selected as preferred (recall that our original procedure selected table 1 as preferred). Similarly, the preferred table for acetaldehyde is that presented on p. 234.

These matrix representations, although explicit, are somewhat cumbersome and, because they are symmetrical, they are wasteful of space. Let us therefore introduce certain conventions* which will provide a more convenient form of representation but which do not alter the results we have already obtained. We shall create a set of four lists, the first of which will show—not the atom whose number corresponds to a particular position in the list—but the number of the lowest numbered node to which it is attached (except for node 1, whose position in the list is left blank). Call this list L1. For example, the L1 list for acetylene would be

Implicit node number	1	2	3	4	3		1		2		4
is attached to	-	1	1	2	H	-	C	≡	C	-	H

which we can write L1 - 1 1 2. The second list contains the actual element symbols corresponding to the nodes whose numbers are implied by the positions of the element symbols in the list. Call this list L2. For acetylene, for example, we now have

L1	-	1	1	2
L2	C	C	H	H

These bonds are given in a third list which is constructed like the first list. Thus, the lists for acetylene would appear

L1	-	1	1	2
L2	C	C	H	H
L3	-	3	1	1

(arrangement) of the nodes. All twenty-four tables in Table 13 are isomorphs of one another.

*These conventions are essentially those introduced by Morgan and employed currently by DuPont and CAS. They are modeled on the "compact list" described by Gluck.

Finally, we may add a fourth list, L4, by means of which ancillary data, such as that describing stereoisomerism, isotopic mass, ionic charge, and unusual valences, can be recorded. This list will be ignored in the remainder of our discussion, even though it is essential for the complete identification of a chemical species.

In order to complete our tabular description of chemical structures using the list formed, we must establish one additional convention, for if we did not, cyclic structures could not be represented. Using cyclohexane as an example, and ignoring hydrogen, we see that it would be numbered

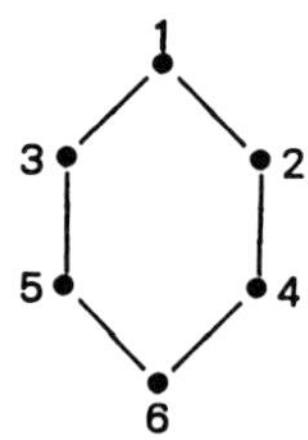

and that the list L1 for the preferred table would appear

L1 - 1 1 2 3 4

from which it is evident that the bond between nodes 5 and 6 has been "left out." This deficiency can be remedied by introducing a sublist of L1 which we call, after Morgan,[22] the ring-closure list (R-C), into which are entered numbered pairs corresponding to the end points of the edges (bonds) which close rings in a structure. The lower number of the pair is entered first, and pairs are entered in ascending lexicographic order. The complete set of lists (exclusive of L4) for cyclohexane is as follows (hydrogen ignored):

							R-C
L1	-	1	1	2	3	4	5-6
L2	C	C	C	C	C	C	
L3	-	1	1	1	1	1	1

In all our early examples, hydrogen atoms were accounted for just as were all other atoms. In contrast, we have ignored hydrogen atoms in illustrations which employed cyclohexane as an example. In fact, it is almost essential to ignore hydrogen if one is to generate and maintain a file of chemical structures represented in connection table form, for the number of hydrogen atoms in a structure is, on the average, greater than the number of other atoms combined. Since we can define a "normal" valence for each element, we can then safely make the assumption that table entries (atoms) with less than the "normal" number of connections are joined with a number of hydrogen atoms equal to the difference between the "normal" valence and the "table" valence. When, however, a table entry has a valence greater than the "normal" valence, an entry must appear in the L4 list which justifies this valence state. In the current CAS connection table, a compromise is employed: the number of hydrogen atoms attached to each noncarbon atom of the table is cited in a separate rank of the table. Typical CAS connection tables are shown in Figure 51.

FILE ORGANIZATION*

We have so far shown how a chemical compound can be represented uniquely and unambiguously by means of a connection table. The type of table described is also highly detailed. But since connection table representations of structure are of value only when large numbers of compounds are to be processed, one may ask whether such bulky representations require large amounts of storage space, and whether by "large" we mean "economically unfeasible." The answer to this question, thus interpreted, is a qualified yes. "Yes," connection tables require considerably more storage than do many linear notations, or chemical names (this is not necessarily bad). "Maybe," the storage requirements make the use of connection tables impractical. Questions of an economic nature can never be answered in an easy way. One must weigh the parameters one sees as significant and judge accordingly. We

*For excellent treatment of this subject, see Lefkovitz,[23] Flores,[24] and D'Imperio.[25]

REG NO 76-15-3
TEXT NS

ATOM NO	1	2	3	4	5	6	7	8
CONN	-	1	1	1	1	2	2	2
ELEMENT	C	C	CL	F	F	F	F	F
BOND		-1	-1	-1	-1	-1	-1	-1
H-COUNT								

STRUCTURE

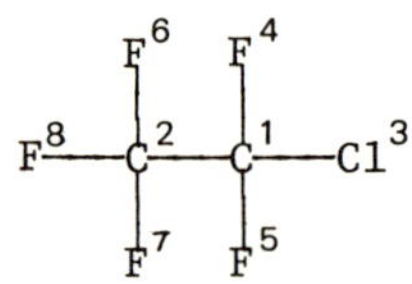

REG NO 12244-67-6
TEXT NS

ATOM NO	1	2	3	4	5	6	7	8	9	10	11	12	13
CONN	-	1	1	1	2	2	3	3	4	5	6	6	9
ELEMENT	C	C	N	N	N	C	C	C	C	C	N	O	CL
BOND		*5	*1	*5	*1	*5	*6	-1	*5	-1	*5	-1	-1
H-COUNT													

RING CLOSURE PAIRS 005 *6 007 009 *5 011

CHARGE 012,-01
DELOCALIZED CHARGE 003,+01 005,+01 007,+01

STRUCTURE

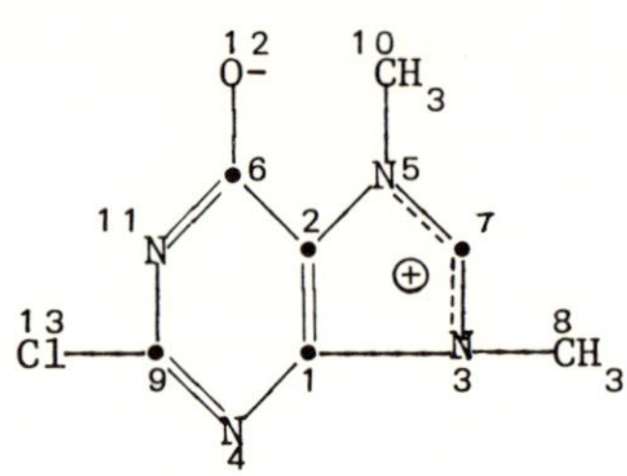

Figure 51–Two Typical CAS Connection Tables
Hyphen = nonring bond; asterisk = ring bond.

shall note here that storage requirements for connection tables are not as bad as one might suppose. Let us consider some of the factors involved in determining storage requirements for a file of connection tables.

The four lists (five, counting the ring-closure sublist) contained in the compact connection table described by Morgan were designed to take advantage of certain characteristics which a large file of chemical structure representations would be expected to exhibit. For instance, there are many six-membered-ring compounds which differ either in bonding or in atoms, or both*. For instance, cyclohexane, cyclohexene, cyclohexadiene (3 isomers possible), benzene, and benzyne are all six-membered carbocycles whose constitutional differences are in bonding. But bonding is displayed, in the type of connection table we have considered, in the third (L3) list. Thus, the first two lists for all seven carbocycles are the same, as Figure 52 illustrates. It seems useless, therefore, to enter these lists into a file every time they occur, when we might find a way of "indexing" the tables to reduce or remove this redundancy. One way to achieve a reduction in redundancy is to enter tables into a file in a way that obviates the repetition of those lists (L1, L2, etc.) which are common to several connection tables. For the carbocycles exemplified above, a technique of file organization which reduces the number of lists to be stored from 24 to 16** is illustrated in Figure 53. A file organized in this way is called a *physically sequential* file. That is, every item in the file must be physically juxtaposed with those items that the file order dictates. Physically sequential file organization is not well-suited to frequent, high-volume updating because the insertion of an item in the file requires that all its successors be moved down in the file to make room for the new item. This disadvantage may be at least partially overcome by resorting to other file organizations. Let us consider one or two alternatives.

*The Ring Index lists over 300 six-membered rings; this number does not include many of the possible variations in unsaturation which such rings may exhibit.[26]

**The technique serves simultaneously for the determination of the existence of a particular table in the file. Note that the L4 lists are not shown in Figures 52 and 53.

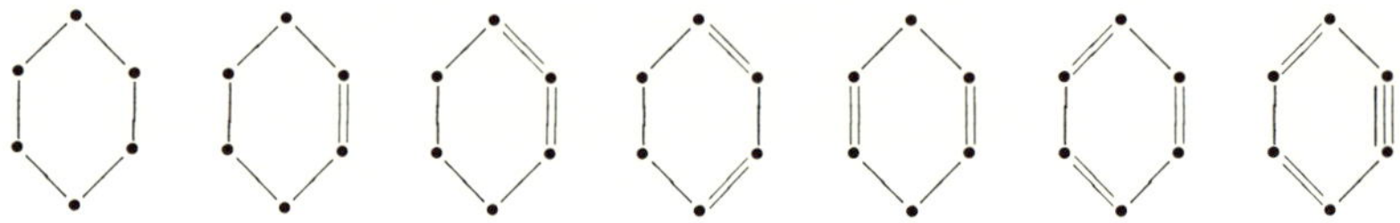

L1	-	1	1	2	3	4	5-6
L2	C	C	C	C	C	C	
L3	-	J	J	J	J	J	J
L1	-	1	1	2	3	4	5-6
L2	C	C	C	C	C	C	
L3	-	J	J	J	J	J	K
L1	-	1	1	2	3	4	5-6
L2	C	C	C	C	C	C	
L3	-	J	J	J	J	K	K
L1	-	1	1	2	3	4	5-6
L2	C	C	C	C	C	C	
L3	-	J	J	J	K	K	J
L1	-	1	1	2	3	4	5-6
L2	C	C	C	C	C	C	
L3	-	J	J	K	K	J	J
L1	-	1	1	2	3	4	5-6
L2	C	C	C	C	C	C	
L3	-	J	K	K	J	J	M
L1	-	1	1	2	3	4	5-6
L2	C	C	C	C	C	C	
L3	-	L	L	L	L	L	L

Figure 52—Lists L1 to L3 for the Seven Carbocycles Shown

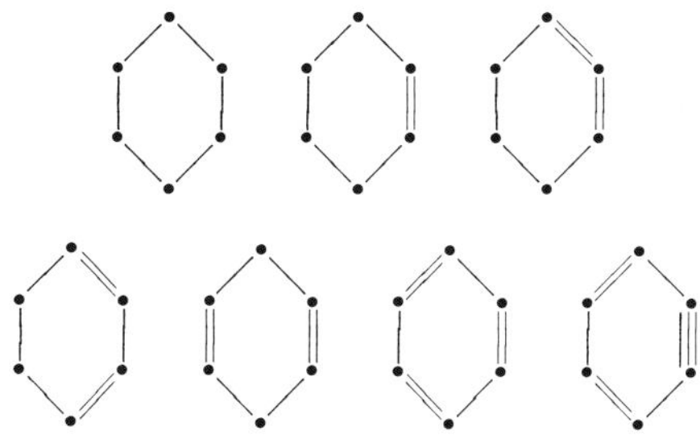

L1	-	1	1	2	3	4	5-6
L2	C	C	C	C	C	C	
L3	-	J	J	J	J	J	J
L3	-	J	J	J	J	J	K
L3	-	J	J	J	J	K	K
L3	-	J	J	J	K	K	J
L3	-	J	J	K	K	J	J
L3	-	J	K	K	J	J	M
L3	-	L	L	L	L	L	L

Figure 53—Lists L1 to L3 for the Seven Carbocycles Organized So That Lists Common to Two or More Structures Are not Repeated (cf. Figure 52)

Suppose that, instead of a single file, four separate files are maintained, each corresponding to one of the four lists of the connection table. Portions of such files, corresponding to the compounds mentioned in Figure 52, are shown in Figure 54. It is easy to see that no space saving has been realized, relative to the physically sequential organization. However, if we introduce a few additional connection tables, it becomes evident (Figure 55) that the new organization requires less space. In fact, in Figure 55, it can be seen that the total number of lists for the 16 compounds represented is 32, in contrast with 39 lists required by physically sequential file organization. Furthermore, the four files may now be organized *time-sequentially* (chronologically) rather than physically sequentially; thus the problem of updating the files is substantially reduced. Nevertheless, the organization just proposed lacks one critical feature: there is no connection between the lists contained in the four files, so one could not reconstitute a connection table once it had been stored. This problem can be resolved by associating with each list a pointer (or pointers) which provide the necessary connections. Several alternative ways of providing the connections are available; only one is suggested here.

With each list we will associate an additional quantity of space which has the following characteristics:

——The space will consist of an integral number, *n,* of subfields of length *m.*

——The value of *n* will vary with the expected distribution of compounds over the various lists, L1, L2, and L3. For L4 lists, *n* will be 3.

——The value of *m* will depend upon the expected number of compounds to be entered into the file.

——(Both *n* and *m* may be periodically adjusted by reorganizing the file.)

A record in each file will have the format shown in Figure 56(a). The four files of Figure 55 would thus appear as illustrated, in part, in Figure 56(b). The accounting data prefixed to each record are useful in searching the file for a particular record. The addition of data to the records, which provide the necessary interfile associations, partly offsets the space savings initially envisioned, but as the collection of compounds grows, this method of organization will show increased space savings as well. In addition, the cost of reorganizing the files is reduced relative to that necessitated by physically sequential file organization. Finally, this file organization is suitable for either sequential- or direct-access processing. It is left to the reader to determine the improvement in performance, relative to physically sequential file organization, to be expected from this method of organization.

File 1	L1	-	1	1	2	3	4		5-6
File 2	L2		C	C	C	C	C	C	
File 3	L3	-	J	J	J	J	J		J
	L3	-	J	J	J	J	J		K
	L3	-	J	J	J	J	K		K
	L3	-	J	J	J	K	K		J
	L3	-	J	J	K	K	J		J
	L3	-	L	L	L	L	L		L
	L3	-	J	K	K	J	J		M
File 4	L4								
	L4								
	L4								
	L4								
	L4								
	L4								
	L4								

Figure 54—Lists L1 to L4 for the Seven Carbocycles of Figure 52, Organized into Four Distinct Files

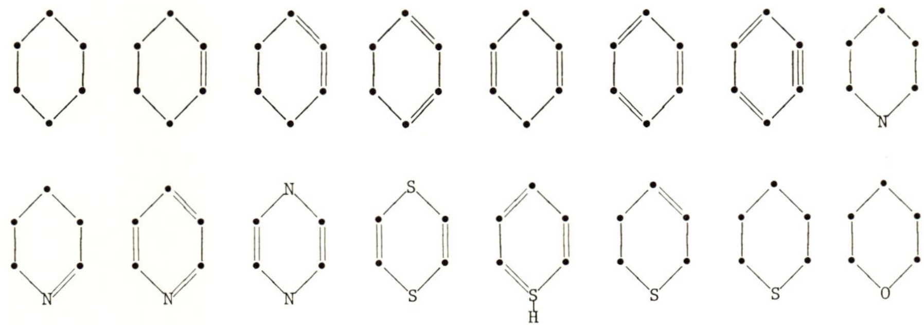

1	File 1	L1	-	1	1	2	3	4	5 - 6
1	File 2	L2	C	C	C	C	C	C	
2		L2	C	C	C	C	C	N	
3		L2	C	C	N	N	C	C	
4		L2	C	C	C	C	C	S	
5		L2	C	C	S	S	C	C	
6		L2	C	C	C	C	C	O	
1	File 3	L3	-	J	J	J	J	J	J
2		L3	-	J	J	J	J	J	K
3		L3	-	J	J	J	J	K	K
4		L3	-	J	J	J	K	K	J
5		L3	-	J	J	K	K	J	J

6		L3	-	K	J	J	J	J	K
7		L3	-	L	L	L	L	L	L
8		L3	-	J	K	K	J	J	M
9		L3	-	J	K	J	J	J	J
1	File 4	L4							
2		L4							
3		L4							
4		L4							
5		L4							
6		L4							
.		.							
.		.							
.		.							
16		L4							

Figure 55—Lists L1 to L4 for the Sixteen Cycles Separated into Four Distinct Files of Unique Entries (no list duplicated)

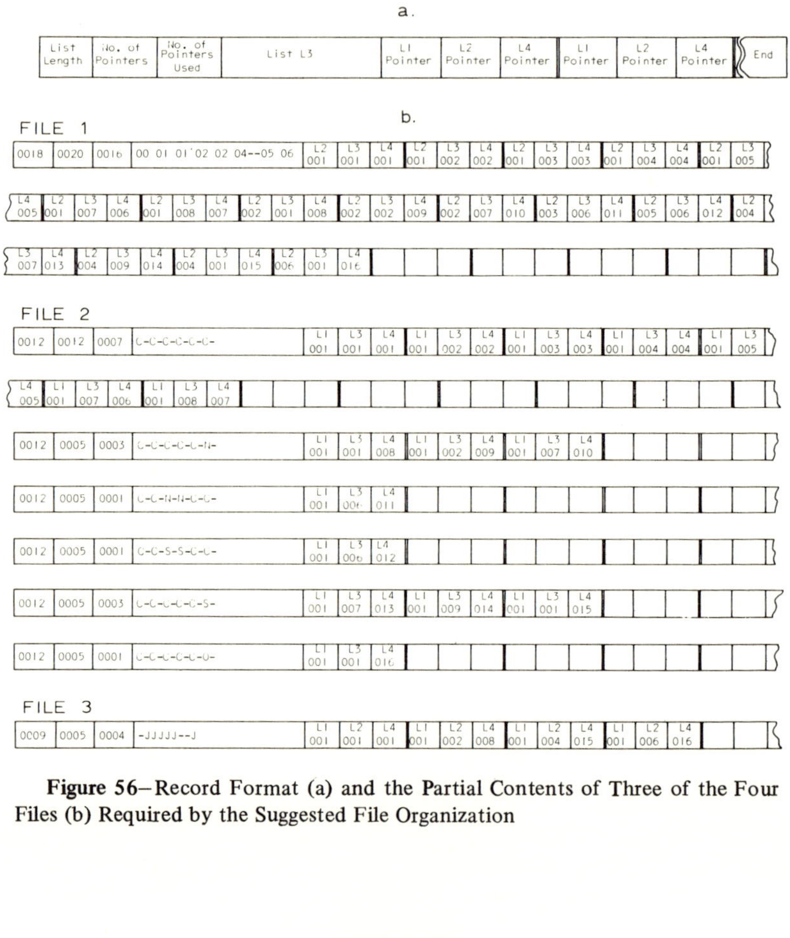

Figure 56–Record Format (a) and the Partial Contents of Three of the Four Files (b) Required by the Suggested File Organization

SUMMARY

In this chapter we have shown how it is possible to represent the topology of a chemical substance by means of a connection table. We have also discussed ways in which unique tables may be consistently derived for each compound entered into a file of connection tables.

Finally, we have briefly considered ways in which files of connection tables may be organized for efficient utilization of storage.

NOTES

1. P. Alexandroff, *Elementary Concepts of Topology* [translated from the German by A. E. Farley] (New York: Dover Publications, Inc., 1961). See also I. Ugi, D. Marquarding, H. Klusacek, G. Gokel, and P. Gillespie, "Chemistry and Logical Structures," *Angewandte Chemie International Edition in English* 9, no. 9 (1970): 703-730; and D. H. Rouvray, "Graph Theory in Chemistry," *Royal Institute of Chemistry Reviews* 4, no. 2 (1971): 173-195.

2. G. W. Wheland, *Syllabus for Advanced Organic Chemistry 321,* Chapters 1-10 of ref. 3 (Chicago: The University of Chicago, 1946).

3. G. W. Wheland, *Advanced Organic Chemistry* (New York: John Wiley and Sons, Inc., 1949), p. 87.

4. C. N. Mooers, "Ciphering Structural Formulas–The Zatopleg System," *Zator Technical Bulletin* 59 (Boston: Zator Co., 79 Milk St., 1951).

5. L. C. Ray and R. A. Kirsch, "Finding Chemical Records by Digital Computers," *Science* 126, no. 3227 (1957): 814-819.

6. E. Meyer and K. Wenke, "Ein System zur topologischen Verschlüsselung organisch-chemischer Strukturformeln für die mechanisierte Dokumentation," *Nachrichten für Dokumentation* 13, no. 1 (1962): 13-19.

7. D. J. Gluck, "A Chemical Structure Storage and Search System Developed at DuPont," *Journal of Chemical Documentation* 5, no. 1 (1965): 43-51. See also "New System Identities and Stores Formulas," *Chemical and Engineering News* 41, no. 49 (1963): 35-36.

8. E. Meyer, "Eine Maschine zur Verschlüsselung chemischer Strukturformeln für die Dokumentation," *Nachrichten für Dokumentation* 13, no. 3 (1962): 144-146; and "Eine topologische Kurzdarstellung chemischer Strukturformeln für die Dokumentation mit elektronischen Rechenanlagen," *Information Storage and Retrieval* 2, no. 4 (1965): 205-215.

9. "Internationales Dokumentationszentrum für Chemie," *Chemie Ingenieur Technik* 39, no. 9-10 (1967): A745-A747; and E. Meyer, "The IDC System for Chemical Documentation," *Journal of Chemical Documentation* 9, no. 2 (1969): 109-113.

10. L. Spialter, "The Atom Connectivity Matrix (ACM) and Its Characteristic Polynomial (ACMCP): A New Computer-Oriented Chemical Nomenclature," *Journal of the American Chemical Society* 85, no. 13 (1963): 2012-2013.

11. G. E. Vleduts and A. L. Seifer, "Machine Information–Logical Methods in Chemistry," *Zavodskaya Laboratoriya* 28, no. 10 (1962): 1224-1232.

12. N. S. Kurnakov, "Introduction to Physico-chemical Analysis," iyd. 3, ONTI 18 (1936).

13. Spialter, "The Atom Connectivity Matrix."

14. A. T. Balaban and F. Harary, "The Characteristic Polynomial Does not Uniquely Determine the Topology of a Molecule," *Journal of Chemical Documentation* 11, no. 4 (1971): 258-259.

15. G. M. Dyson, W. E. Cossum, M. F. Lynch, and H. L. Morgan, "Mechanical Manipulation of Chemical Structure: Molform Computation and Substructure Searching of Organic Structures by the Use of Cipher-Directed, Extended and Random Matrices," *Information Storage and Retrieval* 1, nos. 2-3 (1963): 69-99.

16. Other important efforts to translate from one form of chemical structure representation to another include those of: E. Garfield, "An Algorithm for Translating Chemical Names to Molecular Formulas" (Philadelphia: Institute for Scientific Information, 1961); *Journal of Chemical Documentation* 2, no. 2 (1962): 117-179; G. G. Stetsyura and A. M. Tsukerman, "Automatic Translation of Names of Organic Compounds into Formulas," *Nauchno-Tekniskyia Informatsyia* no. 3 (1962): 17-19; G. G. Vander Stouw, I. Naznitsky, and J. E. Rush, "Procedures for Converting Systematic Names of Organic Compounds into Atom-Bond Connection Tables, *Journal of Chemical Documentation* 7, no. 3 (1967): 165-169; and P. M. Elliott and J. E. Rush, "Translation of Chemical Nomenclature by Syntax-Controlled Techniques," presented before the Chemical Documentation Program, Sixth Middle Atlantic Regional Meeting of the American Chemical Society, Baltimore, February 4-5, 1971.

17. Gluck, "A Chemical Structure Storage and Search System," p. 47.

18. H. L. Morgan, "The Generation of a Unique Machine Description for Chemical Structures–A Technique Developed at Chemical Abstracts Service," *Journal of Chemical Documentation* 5, no. 2 (1965): 107-113.

19. For material on the geometrical aspects of structural representations, see A. P. Terentiev, A. M. Zuckermann, and V. M. Potapov, "Nomenclature of Organic Compounds," Akad.-Nauk. S.S.S.R., Moscow, 1955; A. E. Petrarca, M. F. Lynch, and J. E. Rush, "A Method for Generating Unique Computer Structural Representations of Stereoisomers," *Journal of Chemical Documentation* 7, no. 3 (1967): 154-165; R. S. Cahn, C. K. Ingold, and V. Prelog, *Angewandte Chemie International Edition in English* 5, no. 4 (1966): 385-415; P. M. McDonnell and R. F. Pasternack, "A Line-Formula Notation System for Coordination Compounds," *Journal of Chemical Documentation* 5, no. 1 (1965): 56-60; J. E. Blackwood, C. L. Gladys, K. L. Loening, A. E. Petrarca, and J. E. Rush, "Unambiguous Specification of Stereoisomerism about a Double Bond," *Journal of the American Chemical Society* 90, no. 2 (1968): 509-510; J. E. Blackwood, C. L. Gladys, A. E. Petrarca, W. H. Powell, and J. E. Rush, "Unique and Unambiguous Specification of Stereoisomerism About a Double Bond in Nomenclature and Other Notation Systems," *Journal of Chemical Documentation* 8, no. 1 (1968): 30-32; A. E. Petrarca, M. F. Brown, and J. E. Rush, "A Nomenclature-

Independent Method for Specification of the Stereochemistry of Coordination Compounds," presented before the Division of Inorganic Chemistry, 156th National Meeting of the American Chemical Society, Atlantic City, N.J., September 8-13, 1968; and A. E. Petrarca and J. E. Rush, "Methods for Computer Generation of Unique Configurational Descriptors for Stereoisomeric Square-Planar and Octahedral Complexes," *Journal of Chemical Documentation* 9, no. 1 (1969): 32-37.

20. The history is partly detailed in: *Survey of Chemical Notation Systems,* prepared by the Committee on Chemical Information of the Division of Chemistry and Chemical Technology, National Research Council, and published as National Academy of Sciences Pub. No. 1150, Washington, D.C., 1964; and in *Chemical Structure Information Handling: A Review of the Literature 1962-1968,* prepared by the Committee on Chemical Information of the Division of Chemistry and Chemical Technology, National Research Council, and published as National Academy of Sciences Pub. No. 1733, Washington, D.C., 1969.

21. C. N. Mooers, "Generation of Unique Ciphers for a Finite Network," *Zator Technical Bulletin* 49 (Boston, Mass.: Zator Co., 79 Milk St., 1949); and "Ciphering Structural Formulas."

22. Morgan, "The Generation of a Unique Machine Description for Chemical Structures."

23. D. Lefkovitz, *File Structures for On-Line Systems* (New York: Spartan Books, 1969).

24. I. Flores, *Data Structure and Management* (Englewood Cliffs, N.J.: Prentice-Hall, Inc., 1970).

25. M. E. D'Imperio, "Data Structures and Their Representation in Storage," in M. I. Halpern and C. J. Shaw (eds.), *Annual Review in Automatic Programming,* Vol. 5 (New York: Pergamon Press, 1969), pp. 1-75.

26. A. M. Patterson, L. T. Cappell, and D. F. Walker, *The Ring Index,* 2d ed. (Columbus, Ohio: American Chemical Society, 1960).

Index